FROM BEING TO BECOMING

Pre-Columbian calcite model of a temple, dated before 300 B.C.,
from the state of Guerrero in Mexico. Its height is 10 cm.
Private collection.

FROM BEING
TO BECOMING

TIME AND COMPLEXITY IN
THE PHYSICAL SCIENCES

ILYA PRIGOGINE

Free University of Brussels
and
The University of Texas at Austin

W. H. FREEMAN AND COMPANY
San Francisco

Sponsoring Editor: Peter Renz
Designer: Robert Ishi
Production Coordinator: William Murdock
Illustration Coordinator: Cheryl Nufer
Artists: John and Jean Foster
Compositor: Santype International Limited
Printer and Binder: The Maple-Vail Book Manufacturing Group

Library of Congress Cataloging in Publication Data

Prigogine, Ilya.
    From being to becoming: time and complexity in the
physical sciences.

    Bibliography: p.
    Includes index.
    1. Space and time. 2. Irreversible processes.
3. Physics—Philosophy. I. Title.
QC173.59.S65P76   500.2'01   79-26774
ISBN 0-7167-1107-9
ISBN 0-7167-1108-7 pbk.

Printed in the United States of America

9 8 7 6 5 4 3 2 1

To my colleagues and friends
in Brussels and Austin
whose collaboration has
made this work possible

CONTENTS

Come, press me tenderly upon your breast
But not too hard, for fear the glass might break
This is the way things are: the World
Scarcely suffices for the natural.
But the artificial needs to be confined.

GOETHE, Faust, Part II

PREFACE

This book is about time. I would like to have named it *Time, the Forgotten Dimension*, although such a title might surprise some readers. Is not time incorporated from the start in dynamics, in the study of motion? Is not time the very point of concern of the special theory of relativity? This is certainly true. However, in the dynamical description, be it classical or quantum, time enters only in a quite restricted way, in the sense that these equations are invariant with respect to time inversion, $t \rightarrow -t$. Although a specific type of interaction, the so-called superweak interaction, seems to violate this time symmetry, the violation plays no role in the problems that are the subject of this book.

As early as 1754, d'Alembert noted that time appears in dynamics as a mere "geometrical parameter" (d'Alembert 1754). And Lagrange, more than a hundred years before the work of Einstein and Minkowski, went so far as to call dynamics a four-dimensional geometry (Lagrange 1796). In this view, future and past play the same role. The world lines, the trajectories, followed by the atoms or particles that make up our universe can be traced toward the future or toward the past.

This static view of the world is rooted in the origin of Western science (Sambursky 1963). The Milesian school, of which Thales was one of the most illustrious proponents, introduced the idea of a primordial matter closely related to the concept of conservation of matter. For Thales, a single substance (such as water) forms the primordial matter; all changes in physical phenomena, such as growth and decay, must therefore be mere illusions.

Physicists and chemists know that a description in which past and future play the same role does not apply to all phenomena. Everybody observes that two liquids put into the same vessel generally diffuse toward some homogeneous mixture. In this experiment, the direction of time is essential. We observe progressive homogenization, and the one-sidedness of time becomes evident in the fact that we do not observe spontaneous phase separation of the two mixed liquids. But for a long time such phenomena were excluded from the fundamental description of physics. All time-oriented processes were considered to be the effect of special, "improbable" initial conditions.

At the beginning of this century, this static view was almost unanimously accepted by the scientific community, as will be seen in Chapter 1. But we have since been moving away from it. A dynamical view in which time plays an essential role prevails in nearly all fields of science. The concept of evolution seems to be central to our understanding of the physical universe. It emerged with full force in the nineteenth century. It is remarkable that it appeared almost simultaneously in physics, biology, and sociology, although with quite different specific meanings. In physics it was introduced through the *second law of thermodynamics*, the celebrated law of increase of entropy, which is one of the main subjects of this book.

In the classical view, the second law expressed the increase of molecular disorder; as expressed by Boltzmann, thermodynamic equilibrium corresponds to the state of maximum "probability." However, in biology and sociology, the basic meaning of evolution was just the opposite, describing instead transformations to higher levels of complexity. How can we relate these various meanings of time—time as motion, as in dynamics; time related to irreversibility, as in thermodynamics; time as history, as in biology and sociology? It is evident that this is not an easy matter. Yet, we are living in a single universe. To reach a coherent view of the world of which we are a part, we must find some way to pass from one description to another.

A basic aim of this book is to convey to the reader my conviction that we are in a period of scientific revolution—one in which the very position and meaning of the scientific approach are undergoing reappraisal—a period not unlike the birth of the scientific approach in ancient Greece or of its renaissance in the time of Galileo.

Many interesting and fundamental discoveries have broadened our scientific horizon. To cite only a few: quarks in elementary particle physics; strange objects like pulsars in the sky; the amazing progress of molecular biology. These are landmarks of our times, which are especially rich in important discoveries. However, when I speak of a scientific revolution, I have in mind something different, something perhaps more subtle. Since the beginning of Western science, we have believed in the "simplicity" of the microscopic—molecules, atoms, elementary particles. Irreversibility and evolution appear, then, as illusions related to the complexity of collective behavior of intrinsically simple objects. This conception—historically one of the driving forces of Western science—can hardly be maintained today. The elementary particles that we know are complex objects that can be produced and can decay. If there is simplicity somewhere in physics and chemistry, it is not in the microscopic models. It lies more in idealized macroscopic representations, such as those of simple motions like the harmonic oscillator or the two-body problem. However, if we use such models to describe the behavior of large systems or very small ones, this simplicity is lost. Once we no longer believe in the simplicity of the microscopic, we must reappraise the role of time. We come, then, to the main thesis of this book, which can be formulated as follows:

First, irreversible processes are as *real* as reversible ones; they do not correspond to supplementary approximations that we of necessity superpose upon time-reversible laws.

Second, irreversible processes play a fundamental *constructive* role in the physical world; they are at the bases of important coherent processes that appear with particular clarity on the biological level.

Third, irreversibility is deeply rooted in dynamics. One may say that irreversibility starts where the basic concepts of classical or quantum mechanics (such as trajectories or wave functions) cease to be observables. Irreversibility corresponds not to some supplementary approximation introduced into the laws of dynamics but to an embedding of dynamics within a vaster formalism. Thus, as will be shown, there is a microscopic formulation that extends beyond the conventional formulations of classical and quantum mechanics and *explicitly* displays the role of irreversible processes.

This formulation leads to a unified picture that enables us to relate many aspects of our observations of physical systems to biological ones. The intention is not to "reduce" physics and biology to a single scheme, but to clearly define the various levels of description and to present conditions that permit us to pass from one level to another.

The role of geometrical representations in classical physics is well known. Classical physics is based on Euclidean geometry, and modern developments in relativity and other fields are closely related to extensions of geometrical concepts. But take the other extreme: the field theory used by embryologists to describe the complex phenomena of morphogenesis. It is a striking experience, especially for a nonbiologist, to attend a movie describing the development of, for example, the chicken embryo. We see the progressive organization of a biological space in which every event proceeds at a moment and in a region that make it possible for the process to be coordinated as a whole. This space is functional, not geometrical. The standard geometrical space, the Euclidean space, is invariant with respect to translations or rotations. This is not so in the biological space. In this space the events are processes localized in space and time and not merely trajectories. We are quite close to the Aristotelian view of the cosmos (see Sambursky 1963), which contrasted the world of divine and eternal trajectories with the world of sublunar nature, the description of which was clearly influenced by biological observations.

> The glory, doubtless, of the heavenly bodies fills us with more delight than the contemplation of these lowly things; for the sun and stars are born not, neither do they decay, but are eternal and divine. But the heavens are high and afar off, and of celestial things the knowledge that our senses give is scanty and dim. The living creatures, on the other hand, are at our door, and if we so desire it we may gain ample and certain knowledge of each and all. We take pleasure in the beauty of a statue, shall not the living fill us with delight; and all the more if in the spirit of philosophy we search for causes and recognize the evidence of design. Then will nature's purpose and her deep-seated laws be everywhere revealed, all tending in her multitudinous work to one form or another of the Beautiful.
>
> Aristotle, quoted in Haraway, 1976.

Although the application of Aristotle's biological views to physics has had disastrous consequences, the modern theory of bifurcations and instabilities allows us to see that the two concepts—the geometrical world and the organized, functional world—are not incompatible. This advance will, I think, have a lasting influence.

Belief in the "simplicity" of the microscopic level now belongs to the past. But there is a second reason why I am convinced that we are in the middle of a scientific revolution. The classical, often called "Galilean," view of science was to regard the world as an "object," to try to describe the physical world as if it were being seen from the outside as an object of analysis to which we do not belong. This attitude has been immensely successful in the past. But we have reached the limit of this Galilean view (Koyré 1968). To progress further, we must have a better understanding of our position, the point of view from which we start our description of the physical universe. This does not mean that we must revert to a subjectivistic view of science, but in a sense we must relate knowing to characteristic features of life. Jacques Monod has called living systems "these strange objects," and they are very strange indeed compared with the "nonliving" world (Monod 1970). Thus, one of my objectives is to try to disentangle a few general features of these objects. In molecular biology there has been fundamental progress without which this discussion would not have been possible. But I wish to emphasize other aspects: namely, that living organisms are far-from-equilibrium objects separated by instabilities from the world of equilibrium and that living organisms are necessarily "large," macroscopic objects requiring a coherent state of matter in order to produce the complex biomolecules that make the perpetuation of life possible.

These general characteristics must be incorporated in the answer to the question, What is the meaning of our description of the physical world: that is, from what point of view do we describe it? The answer can only be that we start at a macroscopic level, and all the results of our measurements, even those of the microscopic world, at some point refer back to the macroscopic level. As Bohr emphasized, *primitive concepts* exist; these concepts are not known a priori, but every description must be shown to be compatible with their existence (Bohr 1948). This introduces an element of self-consistency into our description of the physical world. For example, living systems have a sense of the direction of time. Experimentation on even the simplest monocellular organisms reveals that this is so. This direction of time is one of these "primitive concepts." No science—whether of reversible time behavior, as in dynamics, or of irreversible processes—would be possible without it. Therefore one of the most interesting aspects of the theory of dissipative structures developed in Chapters 4 and 5 is that we can now find the roots of this direction of time at the basis of physics and chemistry. In turn, this finding justifies in a self-consistent way the sense of time that we have attributed to ourselves. The concept of time is much more complex than we thought. Time associated with motion was only the first aspect that could be incorporated consistently into the framework of theoretical structures such as classical or quantum mechanics.

We can go further. One of the most striking new results to be described in this book is the appearance of a "second time," a time deeply rooted in fluctuations on the microscopic, dynamical level. This new time is no longer a simple parameter, as in classical or quantum mechanics; rather it is an operator, like those describing quantities in quantum mechanics. Why we need operators to describe the unexpected complexity of the microscopic level is one of the most interesting aspects of the development to be considered in this book.

The recent evolution of science may lead to a better integration of the scientific outlook in the framework of Western culture. There is no doubt that the development of science has, in spite of all its successes, also led to some form of cultural stress (Snow 1964). The existence of "two cultures" is due not only to a lack of mutual curiosity, but also, at least partly, to the fact that the scientific approach has had so little to say about problems, such as time and change, pertaining to literature and art. Although this book does not address problems related to philosophy and human sciences, they are examined by my colleague Isabelle Stengers and myself in another book, *La nouvelle alliance* (Gallimard, 1979), soon to be translated into English. It is interesting to note that there is a strong current both in Europe and in the United States to bring the philosophical and the scientific themes closer together. For example, consider the work of Serres, Moscovici, Morin, and others in France and the recent, provocative article by Robert Brustein, "Drama in the Age of Einstein," published in the *New York Times* on August 7, 1977, in which the role of causality in literature is reappraised.

It is probably not an exaggeration to say that Western civilization is time centered. Is this perhaps related to a basic characteristic of the point of view taken in both the Old and the New Testaments?

It was inevitable that the "timeless" conception of classical physics would clash with the metaphysical conceptions of the Western world. It is not by accident that the entire history of philosophy from Kant through Whitehead was either an attempt to eliminate this difficulty through the introduction of another reality (e.g., the noumenal world of Kant) or a new mode of description in which time and freedom, rather than determinism, would play a fundamental role. Be that as it may, time and change are essential in problems of biology and in sociocultural evolution. In fact, a fascinating aspect of cultural and social changes, in contrast with biological evolution, is the relatively short time in which they take place. Therefore, in a sense, anyone interested in cultural and social matters must consider, in one way or another, the problem of time and the laws of change; perhaps inversely, anyone interested in the problem of time cannot avoid taking an interest in the cultural and social changes of our time as well.

Classical physics, even extended by quantum mechanics and relativity, gave us relatively poor models of time evolution. The deterministic laws of physics, which were at one point the only acceptable laws, today seem like gross simplifications, nearly a caricature of evolution. Both in classical and in quantum mechanics, it seemed that, if the state of a system at a given time were "known" with sufficient accuracy, the future (as well as the past) could at least be predicted in principle. Of course, this is a purely conceptual problem; in fact, we know that we cannot even predict whether it will rain in, say, one month from now. Yet, this theoretical framework seems to indicate that in some sense the present already "contains" the past and the future. We shall see that this is not so. The future is not included in the past. Even in physics, as in sociology, only various possible "scenarios" can be predicted. But it is for this very reason that we are participating in a fascinating adventure in which, in the words of Niels Bohr, we are "both spectators and actors."

The level at which this book has been written is intermediate. Thus, a reader must be familiar with the basic tools of theoretical physics and chemistry. I hope, however, that by adopting this level I can present to a large group of readers a simple introduction to a field that seems to me to have wide implications.

The book is organized in the following way. An introductory chapter is followed by a short survey of what may be called the physics of "being" (classical and quantum mechanics). I emphasize mainly the limits of classical and quantum mechanics to convey to the reader my conviction that, far from being closed, these fields are in a state of rapid development. Indeed, it is only when the simplest problems are considered that our understanding is satisfactory. Unfortunately, many of the popular concepts of the structure of science have as their bases undue extrapolations from these simple situations. Attention is then turned to the physics of "becoming," to thermodynamics in its modern form, to self-organization, and the role of fluctuations. Three chapters deal with the methods now available for building a bridge from being to becoming; they involve kinetic theory and its recent extensions. Only Chapter 8 includes more technical considerations. Readers who do not have the necessary background may turn directly to Chapter 9, in which the main conclusions obtained in Chapter 8 are summarized. Perhaps the most important conclusion is that irreversibility starts where classical and quantum mechanics end. This does not mean that classical and quantum mechanics become wrong; rather they then correspond to idealizations that extend beyond the conceptual possibilities of observation. Trajectories or wave functions have a physical context only if we can give them an observable context, but this is no longer the case when irreversibility becomes part of the physical picture. Thus, the book presents a panorama of problems that may serve as an introduction to a deeper understanding of time and change.

All references to the literature are given at the end of the book. Among them are key references in which the interested reader may find further developments; others are original publications of special interest in the context of this book. The selection is admittedly rather arbitrary and I apologize to the reader for any omissions. Of special relevance is the book by Gregoire Nicolis and myself titled *Self-Organization in Non-equilibrium Systems* (Wiley-Interscience, 1977).

In the preface to the 1959 edition of his *Logic of Scientific Discovery*, Karl Popper wrote: "There is at least one philosophic problem in which all thinking men are interested. It is the problem of cosmology: the problem of understanding the world—including ourselves, and our knowledge, as part of the world." The aim of this book is to show that recent developments in physics and chemistry have made a contribution to the problem so beautifully spelled out by Popper.

As in all significant scientific developments, there is an element of surprise. We expect new insights mainly from the study of elementary particles and from the solutions to cosmological problems. The new surprising feature is that the concept of irreversibility on the intermediate, macroscopic level leads to a revision of the basic tools of physics and chemistry such as classical and quantum mechanics. Irreversibility introduces unexpected features that, when properly understood, give the clue to the transition from being to becoming.

Since the origin of Western science, the problem of time has been a challenge. It was closely associated with the Newtonian revolution and it was the inspiration for Boltzmann's work. The challenge is still with us, but perhaps we are now closer to a more synthetic point of view, which is likely to generate new developments in the future.

I am deeply indebted to my co-workers in Brussels and in Austin for the essential role they have played in helping to formulate and to develop the ideas on which this book is based. Although I cannot thank all of them individually here, I would like to express my gratitude to Dr. Alkis Grecos, Dr. Robert Herman, and Miss Isabelle Stengers for their constructive criticism. I owe special thanks to Dr. Marie Theodosopulu, Dr. Jagdish Mehra, and Dr. Gregoire Nicolis for their constant help in the preparation of the manuscript.

October 1979   *Ilya Prigogine*

Chapter 1

INTRODUCTION:
TIME IN PHYSICS

The Dynamical Description
and Its Limits

In our era, we have made great advances in our knowledge of the natural sciences. The explorable physical world includes a truly fantastic range of dimensions. On the microscopic scale, as in elementary particle physics, we have dimensions of the order of 10^{-22} seconds and 10^{-15} centimeters. On the macroscopic scale, as in cosmology, time can be of the order of 10^{10} years (the age of the universe) and distance of the order of 10^{28} centimeters (the distance to the event horizon; i.e., the furthest distance from which physical signals can be received). Perhaps more important than the enormous range of dimensions over which we can describe the physical world is the recently discovered change in its behavior.

At the beginning of this century, physics seemed to be on the verge of reducing the basic structure of matter to a few stable "elementary particles" such as electrons and protons. We are now far from such a simple description. Whatever the future of theoretical physics may be, "elementary" particles seem to be of such great complexity that the adage concerning "the simplicity of the microscopic" no longer holds.

The change in our point of view is equally valid in astrophysics. Whereas the founders of Western astronomy stressed the regularity and eternal character of celestial motions, such a qualification now applies, at best. to very few, limited aspects such as planetary motion. Instead of finding stability and harmony, wherever we look, we discover evolutionary processes leading to diversification and increasing complexity. This shift in our vision of the physical world leads us to investigate branches of mathematics and theoretical physics that are likely to be of interest in the new context.

For Aristotle, physics was the science of processes, of changes that occur in nature (Ross 1955). However, for Galileo and the other founders of modern physics, the only change that could be expressed in precise mathematical terms was acceleration, the variation in the state of motion. This led finally to the fundamental equation of classical mechanics, which relates acceleration to force, $\mathbf{F}$:

$$m \frac{d^2 \mathbf{r}}{dt^2} = \mathbf{F} \qquad (1.1)$$

Henceforth physical time was identified with the time, t, that appears in the classical equations of motion. We could view the physical world as a collection of trajectories, such as Figure 1.1 shows for a "one dimensional" universe.

A trajectory represents the position $X(t)$ of a test particle as a function of time. The important feature is that dynamics makes no distinction between the future and the past. Equation 1.1 is invariant with respect to the time inversion $t \to -t$: both motions A, "forward" in time, and B, "backward" in time, are possible. However, unless the direction of time is introduced, evolutionary processes cannot be described in any nontrivial way. It is therefore not astonishing that Alexandre Koyré (1968) referred to dynamical motion as "a motion unrelated to time or, more strangely, a motion which proceeds in an intemporal time—a notion as paradoxical as that of a change without change."

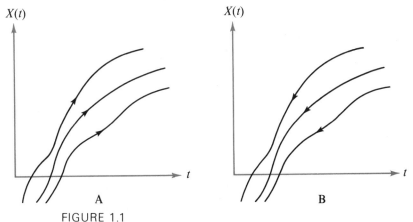

FIGURE 1.1
World lines indicating the time evolution of the coordinate $X(t)$
corresponding to different initial conditions: (A) evolution forward
in time; (B) evolution backward in time.

Again, of the changes that occur in nature, classical physics retained
only motion. Consequently, as Henri Bergson (*Evolution créatice*, 1907;
see Bergson, 1963) and others emphasized, everything is given in classical
physics: change is nothing but a denial of becoming and time is only a
parameter, unaffected by the transformation that it describes. The image
of a stable world, a world that escapes the process of becoming, has
remained until now the very ideal of theoretical physics. The dynamics of
Isaac Newton, completed by his great successors such as Pierre Laplace,
Joseph Lagrange, and Sir William Hamilton, seemed to form a *closed*
universal system, capable of yielding the answer to any question asked.
Almost by definition, a question to which dynamics had no answer was
dismissed as a pseudoproblem. Dynamics thus seemed to give man access
to ultimate reality. In this vision, the rest (including man) appeared only
as a kind of illusion, devoid of fundamental significance.

It thus became the principal aim of physics to identify the microscopic
level to which we could apply dynamics; this microscopic realm could
then serve as the basis for explaining all observable phenomena. Here
classical physics met the program of the Greek atomists, as stated by
Democritus: "Only the atoms and the void."

Today we know that Newtonian dynamics describes only part of our
physical experience; it applies to objects on our own scale whose masses
are measured in grams or tons and whose velocities are much smaller
than that of light. We know that the validity of classical dynamics is limited

by the universal constants, the most important of which are h, Planck's constant, whose value in the cgs system is of the order of 6×10^{-27} erg sec, and c, the velocity of light ($\sim 3 \times 10^{10}$ cm/sec). As the scales of very small objects (atoms, "elementary" particles) or of hyperdense objects (such as neutron stars or black holes) are approached, new phenomena occur. To deal with such phenomena, Newtonian dynamics is replaced by quantum mechanics (which takes into account the finite value of h) and by relativistic dynamics (which includes c). However, these new forms of dynamics—by themselves quite revolutionary—have inherited the idea of Newtonian physics: a static universe, a universe of *being* without *becoming*.

Before further discussion of these concepts, we must ask whether physics can really be identified with some form of dynamics. This question must be qualified. Science is not a closed subject. Examples are the recent discoveries in the field of elementary particles that show how much our theoretical understanding lags behind the available experimental data. But, first, a comment on the role of classical and quantum mechanics in molecular physics, which is the best understood. Can we describe at least qualitatively the main properties of matter in terms of only classical or quantum mechanics? Let us consider in succession certain typical properties of matter. As regards spectroscopic properties, such as emission or absorption of light, there is no doubt that quantum mechanics has been immensely successful in predicting the position of the absorption and emission lines. But with respect to other properties of matter (e.g., the specific heat), we have to go beyond dynamics proper. How does it happen that heating a mole of gaseous hydrogen from, say, 0° to 100°C always requires the same amount of energy if performed at constant volume or constant pressure? Answering this question requires not only knowledge of the structure of the molecules (which can be described by classical or quantum mechanics), but also the assumption that, whatever their histories, any two samples of hydrogen will reach the same "macroscopic" state after some time. We thus perceive a link with the second law of thermodynamics, which is summarized in the next section and which plays an essential role throughout this book.

The role of nondynamical elements becomes even greater when non-equilibrium properties, such as viscosity and diffusion, are included. To calculate such coefficients, we must introduce some form of kinetic theory or a formalism involving a "master equation" (see Chapter 7). The details of the calculation are not important. The main point is that, in addition to the tools provided by classical or quantum dynamics, we need supplementary tools, which will be described briefly before investigating their position with respect to dynamics. Here we encounter the main subject of this book: the role of time in the description of the physical universe.

The Second Law of Thermodynamics

As already mentioned, dynamics describes processes in which the direction of time does not matter. Clearly, there are other situations in which this direction does indeed play an essential role. If we heat part of a macroscopic body and then isolate this body thermally, we observe that the temperature gradually becomes uniform. In such processes, then, time displays an obvious "one-sidedness." Engineers and physical chemists have given such processes extensive study since the end of the eighteenth century. The second law of thermodynamics as formulated by Rudolf Clausius (see Planck, 1930) strikingly summarizes their characteristic features. Clausius considered isolated systems, which exchange neither energy nor matter with the outside world. The second law then implies the existence of a function S, the entropy, which increases monotonically until it reaches its maximum value at the state of thermodynamic equilibrium:

$$\frac{dS}{dt} \geq 0 \qquad (1.2)$$

This formulation can be easily extended to systems that exchange energy and matter with the outside world (see Figure 1.2).

We must distinguish two terms in the entropy change, dS: the first, $d_e S$, is the transfer of entropy across the boundaries of the system; the second, $d_i S$, is the entropy produced within the system. According to the second law, the entropy production inside the system is positive:

$$dS = d_e S + d_i S, \qquad d_i S \geq 0 \qquad (1.3)$$

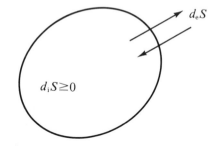

FIGURE 1.2
An open system in which $d_i S$ represents
entropy production and $d_e S$ represents
entropy exchange between system
and environment.

$d_i S \geq 0$

$d_e S$

It is in this formulation that the basic distinction between reversible and irreversible processes becomes essential. Only irreversible processes contribute to entropy production. Examples of irreversible processes are chemical reactions, heat conduction, and diffusion. On the other hand, reversible processes may correspond to wave propagation in the limit in which the absorption of the wave is neglected. The second law of thermodynamics, then, states that irreversible processes lead to a kind of one-sidedness of time. The positive time direction is associated with the increase of entropy. Let us emphasize how strongly and specifically the one-sidedness of time appears in the second law. It postulates the existence of a function having quite specific properties such that in an isolated system it can only increase in time. Such functions play an important role in the modern theory of stability initiated by Aleksander Lyapounov's classic work. (References can be found in Nicolis and Prigogine, 1977.)

There are other instances of the one-sidedness of time. For example, in the superweak interaction, the equations of dynamics do not admit the inversion $t \to -t$. But they are *weaker* forms of one-sidedness; they can be accommodated in the framework of the dynamical description and do not correspond to irreversible processes as introduced by the second law.

Because we shall concentrate on processes that lead to Lyapounov functions, this concept must be examined in more detail. Consider a system whose evolution is described by some variables X_i, which may, for example, represent concentrations of chemical species. The evolution of such a system may be given by rate equations of the form

$$\frac{dX_i}{dt} = F_i(\{X_i\}) \tag{1.4}$$

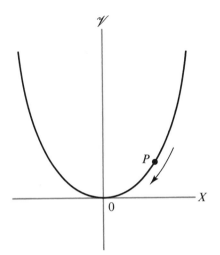

FIGURE 1.3
The concept of asymptotic stability: if a perturbation leads to point P, the system will respond through an evolution leading back to the equilibrium point 0.

in which F_i is the overall production rate of the component X_i; there is one equation for each component (examples are given in Chapters 4 and 5). Suppose that, for $X_i = 0$, all the reaction rates vanish. This is then an *equilibrium point* for the system. We may now ask, If we start with nonvanishing values of the concentrations, X_i, will this system evolve toward the equilibrium point $X_i = 0$? In today's terminology, is the state $X_i = 0$ an *attractor*? Lyapounov functions enable us to tackle this problem. We consider a function of the concentrations, $\mathscr{V} = \mathscr{V}(X_1, \ldots, X_n)$, and we suppose that it is *positive* throughout the region of interest and vanishes at $X = 0$.* We then consider how $\mathscr{V}(X_1, \ldots, X_n)$ varies as the concentrations, X_i, evolve. The time derivative of this function as the concentrations evolve according to the rate equations (1.4) is:

$$\frac{d\mathscr{V}}{dt} = \sum_i \frac{\partial \mathscr{V}}{\partial X_i} \frac{dX_i}{dt} \tag{1.5}$$

Lyapounov's theorem asserts that the equilibrium state will be an attractor if $d\mathscr{V}/dt$, the time derivative of $\mathscr{V}$, has the opposite sign of $\mathscr{V}$; that is, if derivative in our example is negative. The geometrical meaning of this condition is evident; see Figure 1.3. For isolated systems, the second

---

* In general, a Lyapounov function may also be negative definite, but its first derivative must be positive definite (see, e.g., equation 4.28).

law of thermodynamics states that a Lyapounov function exists and that, for such systems, thermodynamic equilibrium is an attractor of non-equilibrium states. This important point can be illustrated by a simple problem in heat conductivity. The time change of temperature T is described by the classical Fourier equation:

$$\frac{\partial T}{\partial t} = \kappa \frac{\partial^2 T}{\partial x^2} \tag{1.6}$$

in which κ is the heat conductivity $(\kappa > 0)$. A Lyapounov function for this problem can easily be found. We can take, for example,

$$\Theta(T) = \int \left(\frac{\partial T}{\partial x}\right)^2 dx \geq 0 \tag{1.7}$$

It can be verified immediately that, for fixed boundary conditions,

$$\frac{d\Theta}{dt} = -2\kappa \int \left(\frac{\partial^2 T}{\partial x^2}\right)^2 dx \leq 0 \tag{1.8}$$

and the Lyapounov function $\Theta(T)$ decreases indeed to its minimum value when thermal equilibrium is reached. Inversely, the uniform temperature distribution is an attractor for initial nonuniform distributions.

Max Planck emphasized, quite rightly, that the second law of thermodynamics distinguishes between various types of states in nature, some of which act as attractors for others. Irreversibility is the expression of this *attraction* (Planck 1930).

Such a description of nature is clearly very different from the dynamical description: two different initial temperature distributions reach the same uniform distribution in time (see Figure 1.4). The system possesses an intrinsic "forgetting" mechanism. How different this is from the dynamical "world line" view, in which the system always follows a given trajectory. There is a theorem in dynamics that shows that two trajectories can never cross; at most they may meet asymptotically (for $t \to \pm\infty$) at singular points.

Let us now briefly consider how irreversible processes can be described in terms of molecular events.

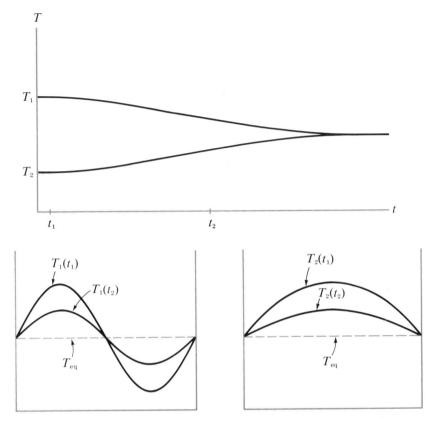

FIGURE 1.4
Approach to thermal equilibrium. Different initial distributions
such as T_1 and T_2 lead to the same temperature distribution.

Molecular Description
of Irreversible Processes

Let us first ask what an increase in entropy means in terms of the
molecules involved. To find an answer, we must explore the microscopic
meaning of entropy. Ludwig Boltzmann, the first to note that entropy is a
measure of molecular disorder, concluded that the law of entropy increase is simply a law of increasing disorganization. Consider, for exam-

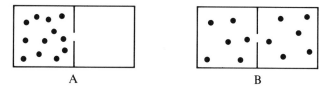

FIGURE 1.5
Two different distributions of molecules between two compartments: (A) $N = N_1 = 12$, $N_2 = 0$; (B) $N_1 = N_2 = 6$. After a sufficient lapse of time, distribution B represents the most probable configuration, the analogue of thermodynamic equilibrium.

ple, a container partitioned into two equal volumes (see Figure 1.5). The number of ways, P, in which N molecules can be divided into two groups, N_1 and N_2, is given by the simple combinatorial formula

$$P = \frac{N!}{N_1! N_2!} \tag{1.9}$$

in which $N! = N(N - 1)(N - 2) \cdots 3 \cdot 2 \cdot 1$. The quantity P is called the *number of complexions* (see Landau and Lifschitz, 1968).

Starting from any initial values of N_1 and N_2 we may perform a simple experiment, a "game" proposed by Paul and Tatiana Ehrenfest to illustrate Boltzmann's ideas (for more details see Eigen and Winkler, 1975). We choose a particle at random and agree that, when chosen, it will change its compartment. As could be expected, after a sufficiently long time an equilibrium is reached in which, except for small fluctuations, there is an equal number of molecules in the two compartments ($N_1 \simeq N_2 \simeq N/2$).

It can be easily seen that this situation corresponds to the maximum value of P and in the course of evolution P increases. Thus Boltzmann identified the number of complexions, P, with the entropy through the relation

$$S = k \log P \tag{1.10}$$

in which k is Boltzmann's universal constant: an entropy increase expresses growing molecular disorder, as indicated by the increasing number of complexions. In such an evolution, the initial conditions are "forgotten." Whenever one compartment is at first favored with more

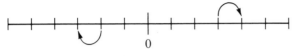

FIGURE 1.6
A one-dimensional random walk.

particles than the other compartment, this lack of symmetry is always destroyed in time.

If P is associated with the "probability" of a state as measured by the number of complexions, then the increase of entropy corresponds to the evolution toward the "most probable" state. We shall return to this interpretation later. It was through the molecular interpretation of irreversibility that the concept of probability first entered theoretical physics. This was a decisive step in the history of modern physics.

We can push such probability arguments still further to obtain quantitative formulations that describe how irreversible processes evolve with time. Consider, for example, the well-known random walk problem, an idealized but nevertheless successful model for Brownian motion. In the simplest example, a one-dimensional random walk, a molecule makes a one-step transition at regular time intervals (see Figure 1.6). With the molecule initially at the origin, we ask for the probability of finding it at point m, after N steps. If the probability that the molecule proceeds forward or backward is assumed to be one-half, we find that

$$W(m, N) = \left(\frac{1}{2}\right)\frac{N!}{[\frac{1}{2}(N + m)]! \, [\frac{1}{2}(N - m)]!} \tag{1.11}$$

Thus, to arrive at point m after N steps, some $\frac{1}{2}(N + m)$ steps must be taken to the right and some $\frac{1}{2}(N - m)$ to the left. Equation 1.11 gives the number of such distinct sequences multiplied by the overall probability of an arbitrary sequence of N steps. (For details, see Chandrasekhar, 1943.)

Expanding the factorials, we obtain the asymptotic formula corresponding to a Gaussian distribution:

$$W(m, N) = \left(\frac{2}{\pi N}\right)^{1/2} e^{-m^2/2N} \tag{1.12}$$

Using the notation $D = \frac{1}{2}nl^2$, in which l is the distance between two sites and n the number of displacements per unit time, this result can be written:

$$W(x, t) = \frac{1}{2(\pi Dt)^{1/2}}\, e^{-x^2/4Dt} \tag{1.13}$$

in which $x = ml$. This is the solution of a one-dimensional diffusion equation identical in form to the Fourier equation (equation 1.6, but κ is replaced by D). Evidently, this is a very simple example; in Chapter 7, consideration is given to more sophisticated techniques for deriving irreversible processes from kinetic theory. Here, however, we may ask the fundamental questions, What is the position of irreversible processes in our description of the physical world? What is the relation of these processes to dynamics?

Time and Dynamics

In classical and quantum dynamics, the fundamental laws of physics are taken to be symmetrical in time. Thermodynamic irreversibility corresponds to some kind of approximation added to dynamics. An often quoted example was given by Josiah Gibbs (1902): if we put a drop of black ink into water and stir it, the medium will look gray. This process would seem to be irreversible. But if we could follow each molecule we would recognize that in the microscopic realm the system has remained heterogeneous. Irreversibility would be an illusion caused by the observer's imperfect sense organs. It is true that the system has remained heterogeneous, but the scale of heterogeneity, initially macroscopic, has become microscopic. The view that irreversibility is an illusion has been very influential and many scientists have tried to tie this illusion to mathematical procedures, such as coarse graining, that would lead to irreversible processes. Others with similar aims have tried to work out the conditions of macroscopic observation. None of these attempts has led to conclusive results.

It is difficult to believe that the observed irreversible processes, such as viscosity, decay of unstable particles, and so forth, are simply illusions caused by lack of knowledge or by incomplete observation. Because we know the initial conditions even in simple dynamical motion only approximately, future states of motion become more difficult to predict as time increases. Still, it does not seem meaningful to apply the second law of thermodynamics to such systems. Properties like specific heat and compressibility, which are closely associated with the second law, are meaningful for a gas formed by many interacting particles but are meaningless when applied to such simple dynamical systems as the planetary system. Therefore, irreversibility must have some basic connection with the dynamical nature of the system.

The opposite notion has also been considered: perhaps dynamics is incomplete; perhaps it should be expanded to include irreversible processes. This attitude is also difficult to maintain, because for simple types of dynamical systems the predictions, both of classical and quantum mechanics, have been remarkably well verified. It is enough to mention the success of space travel, which requires very accurate computation of the dynamical trajectories.

In recent times it has been repeatedly asked whether quantum mechanics is complete in connection with the so-called measurement problem (to which we return in Chapter 7). It has even been suggested that, to include the irreversibility of the measurement, new terms would have to be added to the Schrödinger equation describing the dynamics of quantum systems (see Chapter 3).

We come here to the very formulation of the subject of this book. Using the philosophical vocabulary, we can relate the "static" dynamical description with *being*; then the thermodynamic description, with its emphasis on irreversibility, can be related to *becoming*. The aim of this book, then, is to discuss the relation between the physics of being and the physics of becoming.

Before that relationship can be dealt with, however, the physics of being must be described. This is done by means of a short outline of classical and quantum mechanics, emphasizing their basic concepts and their present limitations. Then the physics of becoming is addressed, with a short presentation of modern thermodynamics, including the basic problem of self-organization.

We are then ready to examine our central problem: the transition between being and becoming. To what extent can we present today a logically coherent, though necessarily incomplete, description of the physical world? Have we reached some unity of knowledge or is science broken into various parts based on contradictory premises? Such questions will lead us to a deeper understanding of the role of time. The problems of unity of science and of time are so intimately connected that we cannot treat one without the other.

Part I

THE PHYSICS
OF BEING

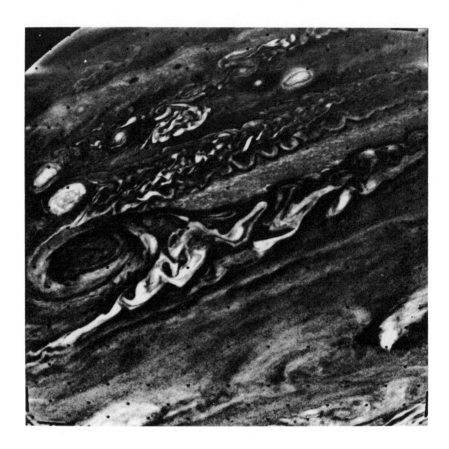

EMERGENCE OF ORDER IN FLUID FLOWS

The ordered structures of a storm arise from complex nonlinear interactions in fluid systems far from equilibrium. The photograph above is of large-scale eddies in Jupiter's atmosphere.

Nonlinear interactions also lead to the emergence of paired vortices at the boundary between two layers of fluid flowing at different velocities, as shown in the computer-drawn graphs on the facing page. Lines of equal vorticity have been plotted.

Initially, the mixing layer is turbulent and has only small-scale structure. By computer simulation, Ralph Metcalf and James Riley show how small perturbations of the mixing layer evolve into various types of large-scale vortices. These simulations closely match experimental work

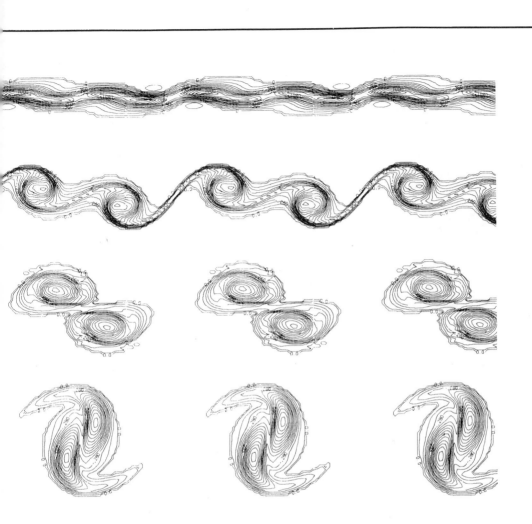

done on mixing layers. Whether flows dissipate in turbulent chaos or lead
to large-scale order depends on the existence and nature of instabilities
in the system.

The photograph of Jupiter is reproduced through the courtesy of the
National Aeronautics and Space Administration and the computer plots through
the courtesy of James Riley and Ralph Metcalfe. Further information
about the computer simulation is given in a paper titled "Direct Numerical
Simulation of a Perturbed, Turbulent Mixing Layer," AIAA-80-0274
(presented at the AIAA 18th Aerospace Sciences Meeting, Pasadena,
California, January 14–16, 1980), which can be obtained from Flow
Research Company, Kent, Washington 98031.

Chapter 2

CLASSICAL DYNAMICS

Introduction

Classical dynamics is the oldest part of present-day theoretical physics. It might even be said that modern science began when Galileo and Newton formulated dynamics. A number of the greatest scientists of Western civilization, such as Lagrange, Hamilton and Henri Poincaré, have made decisive contributions to classical dynamics; moreover, classical dynamics was the starting point of the scientific revolutions of the twentieth century, such as relativity and quantum theory.

Unfortunately, most college and university textbooks present classical dynamics as if it were a closed subject. We shall see that it is not. In fact, it is a subject in rapid evolution. In the past twenty years, Andrei Kolmogoroff, Vladimir Arnol'd, and Jürgen Moser, among others, have introduced important new insights, and further developments can be expected in the near future (see Moser, 1972).

Classical dynamics has been the prototype of the scientific approach. In French the term "rational" mechanics is often used, implying that the laws of classical mechanics are the very laws of reason. Among the characteristics attributed to classical dynamics was that of strict determinism. In dynamics a fundamental distinction is made between initial conditions, which may be given arbitrarily, and the equations of motion, from which the system's later (or earlier) dynamic state can be calculated. As will be seen, this belief in strict determinism is justified only when the notion of a well-defined initial state does not correspond to an excessive idealization. Modern dynamics was born with Johannes Kepler's laws of planetary motion and with Newton's solution of the "two body" problem. However, dynamics becomes enormously more complicated as soon as we take into account a third body—a second planet, for instance. If the system is sufficiently complex (as in the "three body" problem), knowledge (of whatever finite precision) of the system's initial state generally does not allow us to predict the behavior of this system over long periods of time. This uncertainty persists even when precision in the determination of the initial state becomes arbitrarily large. It becomes impossible, even in principle, to know whether, for instance, the solar system that we inhabit is stable for *all* future times. Such considerations considerably limit the usefulness of the concept of trajectories or world lines. We must, then, consider *ensembles* of world lines compatible with our measurements (see Figure 2.1). But once we leave the consideration of single trajectories, we leave the model of strict determinism. We can make only statistical predictions, forecasting *average* results.

It is a curious turn of events. For years, the proponents of classical orthodoxy have tried to free quantum mechanics from its statistical aspects (see Chapter 3): Albert Einstein's remark is well known: "God does not play dice." Now we see that, when long periods of time are considered, classical dynamics itself needs statistical methods. More important still, even classical dynamics, perhaps the most elaborated of all theoretical sciences, is not a "closed" science: we can pose meaningful questions to which it yields no answers.

Because classical dynamics is the oldest of all theoretical sciences, its development illustrates, in many ways, dynamics of the evolution of science. We can see the birth of *paradigms*, their growth and decay.

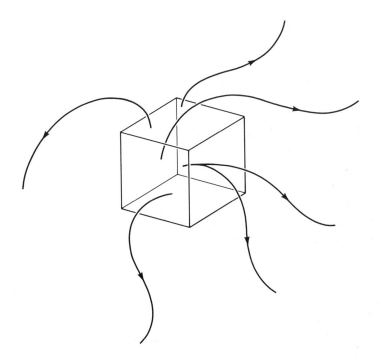

FIGURE 2.1
Trajectories originating from a finite
region in phase space corresponding to the
initial state of the system.

Examples of such paradigms are the concepts of *integrable* and *ergodic* dynamic systems that will be described in the next sections of this chapter. Of course, no systematic description of the theoretical basis of classical dynamics could be presented in this chapter; we can only emphasize certain relevant features.

Hamiltonian Equations of Motion and Ensemble Theory

In classical mechanics it is convenient to describe the state of a system of point particles by the coordinates $q_1, \ldots, q_s$ and momenta $p_1, \ldots, p_s$. Of utmost importance is the energy of the system when expressed in terms of these variables. It generally takes the form

$$H = E_{\text{kin}}(p_1, \ldots, p_s) + V_{\text{pot}}(q_1, \ldots, q_s) \qquad (2.1)$$

in which the first part depends only on the momenta and is the kinetic energy, and the second part depends on the coordinates and is the potential energy (for details, see Goldstein, 1950). The energy expressed in these variables is the *Hamiltonian*, which plays a central role in classical dynamics. In this discussion, only conservative "systems" in which H does not depend explicitly on time are considered. A simple example would be a one-dimensional harmonic oscillator for which the Hamiltonian is

$$H = \frac{p^2}{2m} + k\frac{q^2}{2} \qquad (2.2)$$

in which m is the mass and k is the spring constant related to the frequency v (or angular velocity ω) of the oscillator by

$$v = \frac{1}{2\pi}\left(\frac{k}{m}\right)^{1/2} \quad \text{or} \quad \omega \equiv 2\pi v = \left(\frac{k}{m}\right)^{1/2} \qquad (2.3)$$

In many-body systems, the potential energy is often the sum of two-body interactions, as in gravitational or electrostatic systems.

The central point for us is that, once the Hamiltonian H is known, the motion of the system is determined. Indeed, the laws of classical dynamics may be expressed in terms of Hamilton's equations:

$$\frac{dq_i}{dt} = \frac{\partial H}{\partial p_i} \quad \text{and} \quad \frac{dp_i}{dt} = -\frac{\partial H}{\partial q_i} \qquad (i = 1, 2, \ldots, s) \qquad (2.4)$$

A great achievement of classical dynamics is that its laws can be expressed in terms of a single quantity, the Hamiltonian.

Imagine a space of $2s$ dimensions whose points are determined by the coordinates $q_1, \ldots, p_s$. This space is called the *phase space*. To each mechanical state there corresponds a point P_t of this space. The position of the initial point P at time t_0, together with the Hamiltonian, completely determines the evolution of the system.

Let us consider an arbitrary function of $q_1, \ldots, p_s$. Employing Hamilton's equations (2.4), its change with time will be given by:

$$\frac{df}{dt} = \sum_{i=1}^{s} \left[\frac{\partial f}{\partial q_i} \frac{dq_i}{dt} + \frac{\partial f}{\partial p_i} \frac{dp_i}{dt} \right]$$

$$= \sum_{i=1}^{s} \left[\frac{\partial f}{\partial q_i} \frac{\partial H}{\partial p_i} - \frac{\partial f}{\partial p_i} \frac{\partial H}{\partial q_i} \right] \equiv [f, H] \qquad (2.5)$$

in which $[f, H]$ is called the *Poisson bracket* of f with H. The condition for the invariance of f is, therefore,

$$[f, H] = 0 \qquad (2.6)$$

Clearly,

$$[H, H] = \sum_{i=1}^{s} \left[\frac{\partial H}{\partial p_i} \frac{\partial H}{\partial q_i} - \frac{\partial H}{\partial q_i} \frac{\partial H}{\partial p_i} \right] = 0 \qquad (2.7)$$

This relation expresses simply the conservation of energy.

To make the connection between dynamics and thermodynamics, it is very useful to introduce, as did Gibbs and Einstein, the idea of a *representative ensemble* (see Tolman, 1938). Gibbs has defined it as follows: "We may imagine a great number of systems of the same nature, but differing in the configurations and velocities which they have at a given instant, and differing not merely infinitesimally, but it may be so as to embrace every conceivable combination of configurations and velocities ...."

The basic idea, therefore, is that instead of considering a single dynamic system, we consider a collection of systems, all corresponding to the same Hamiltonian. The choice of this collection, or ensemble, depends on the conditions imposed on the systems (we may for example consider isolated systems or systems in contact with a thermostat) and on our knowledge of the initial conditions. If the initial conditions are well defined, the ensemble will be sharply concentrated in some region of phase space; if they are poorly defined, the ensemble will be distributed over a wide region in phase space.

For Gibbs and Einstein, the ensemble point of view was merely a convenient computational tool for calculating average values when exact

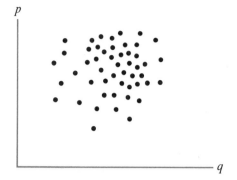

FIGURE 2.2
Gibbs ensemble. Systems, whose state is described by the
various points, have the same Hamiltonian and are subject to the
same constraints, but differ in their initial conditions.

initial conditions were not prescribed. As will be seen in this chapter, as
well as in Chapter 7, the importance of the ensemble point of view goes
much further than originally conceived by Gibbs and Einstein.

A Gibbsian ensemble of systems can be represented by a "cloud" of
points in the phase space (see Figure 2.2). In the limit in which each
region contains a large number of points, the "cloud" can be described
as a continuous fluid with a density of

$$\rho(q_1, \ldots, q_s, p_1, \ldots, p_s, t) \tag{2.8}$$

in phase space. Because the number of points in the ensemble is arbitrary,
ρ will be normalized; that is,

$$\int \rho(q_1, \ldots, q_s, p_1, \ldots, p_s, t) \, dq_1, \ldots, dp_s = 1 \tag{2.9}$$

Therefore

$$\rho \, dq_1, \ldots, dp_s \tag{2.10}$$

represents the *probability* of finding at time t a representative point in the
volume element $dq_1, \ldots, dp_s$ of phase space.

The change of density in every volume element of phase space is due to
the difference of the flows across its boundaries. The remarkable feature is
that the flow in phase space is "incompressible." In other words, the
divergence of the flow vanishes. Indeed, using Hamilton's equations (2.4),

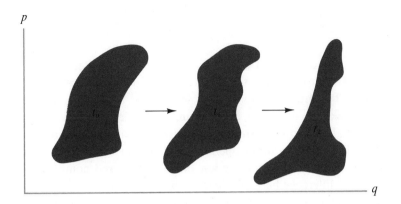

FIGURE 2.3
Preservation of volume in phase space.

we have

$$\sum_{i=1}^{s} \left[\frac{\partial}{\partial q_i} \left(\frac{dq_i}{dt} \right) + \frac{\partial}{\partial p_i} \left(\frac{dp_i}{dt} \right) \right] = 0 \qquad (2.11)$$

As a result, the volume in phase space is preserved in time (see Figure 2.3).

Using equation 2.11, we obtain a simple equation of motion for the phase-space density ρ. As is shown in all textbooks (Tolman 1938), this is the well-known Liouville equation, which can be written in the form

$$\frac{\partial \rho}{\partial t} = - \sum_{i=1}^{s} \left[\frac{\partial H}{\partial p_i} \frac{\partial \rho}{\partial q_i} - \frac{\partial H}{\partial q_i} \frac{\partial \rho}{\partial p_i} \right]$$

$$= [H, \rho] \qquad (2.11')$$

in which (as it is in equation 2.5) the bracket is the Poisson bracket of H with ρ. As it is often convenient to use an *operator formulation*, we simply multiply equation 2.11 by $i = \sqrt{-1}$ and write

$$i \frac{\partial \rho}{\partial t} = L\rho \qquad (2.12)$$

in which L is now the linear *operator*:

$$L = -i \frac{\partial H}{\partial p} \frac{\partial}{\partial q} + i \frac{\partial H}{\partial q} \frac{\partial}{\partial p} \qquad (2.13)$$

The concept of an operator is discussed in greater detail in the next section. To simplify the notations, we have considered a single degree of freedom. The multiplication by i is introduced to make L a *Hermitian operator* like the operators of quantum mechanics studied in Chapter 3. The formal definition of Hermitian operators can be found in any textbook. The definition of an operator in quantum mechanical systems is given in Chapter 3, in the section on operators and complementarity. The basic difference between them lies in the *space* in which they act: in classical dynamics L acts in the *phase* space, whereas in quantum mechanics the operators act in the *coordinate* space or in the momentum space. The Liouville operator has been used extensively in recent work in statistical mechanics (see Prigogine, 1962).

Our interest in ensemble theory is obvious. Even if we do not know the exact initial conditions, we may consider the Gibbs density and calculate the average value of any mechanical property $A(p, q)$ such as

$$\langle A \rangle = \int A(p, q)\rho \, dq \, dp \tag{2.14}$$

using the ensemble average.

Note also that it is easy to give the formal solution of the Liouville equation 2.12 as

$$\rho(t) = e^{-iLt}\rho(0) \tag{2.12$'$}$$

This expression can be verified by straightforward differentiation. A word of caution is necessary here. The Gibbs ensemble approach introduces the probability concept through the density function ρ in phase space. This allows the study of both *pure* cases, for which the initial conditions are prescribed, and *mixtures*, corresponding to various possible initial conditions. In any case, the time evolution of the density function has a strictly deterministic dynamical character. There is *no simple connection* with *probabilistic (or "stochastic") processes*, such as Brownian motion, which is described in Chapter 1. Concepts such as transition probabilities do not appear here. One striking difference is in the role of time. Solution 2.12$'$ is valid for all values of t positive and negative, whereas solution 1.13 refers only to t positive. (In mathematical terms, solution 2.12$'$ corresponds to a group and 2.13 to a semigroup.)

Operators

Operators are generally introduced in connection with quantum mechanics. The quantum mechanical aspects are discussed in Chapter 3, but for now it is sufficient merely to emphasize that operators also appear in classical dynamics when the ensemble point of view is adopted. Indeed, the concept of the Liouville operator has already been introduced in equation 2.13.

In general, an operator has *eigenfunctions* and *eigenvalues*. When an operator acts on one of its eigenfunctions, the result is the eigenfunction multiplied by its associated eigenvalue. Consider, for example, the operator A corresponding to second-order differentiation:

$$A \equiv \frac{d^2}{dx^2} \tag{2.15}$$

If it acts on an arbitrary function (say x^2), the operator changes that function into another one. However, certain functions are left unchanged: for example, consider the "eigenvalue problem"

$$\frac{d^2}{dx^2} u = \lambda u \tag{2.16}$$

having the solutions

$$u = \sin kx \tag{2.17}$$

and

$$\lambda = -k^2 \tag{2.18}$$

in which k is a real number. These are the eigenfunctions and the eigenvalues, respectively, associated with the operator.

The eigenvalues may be either discrete or continuous. To understand this difference, let us reconsider the eigenvalue problem (2.16). So far, boundary conditions have not been introduced, but we now impose the condition that the eigenfunction be zero at the boundaries of the domain corresponding to $x = 0$ and $x = L$. These are the boundary conditions that arise naturally in quantum mechanics. Their physical interpretation is that the particle is trapped inside this domain. It is easy to satisfy these boundary conditions. Indeed the conditions

$$\sin kx = 0 \quad \text{for } x = 0, L \qquad (2.19)$$

lead to

$$kL = n\pi \qquad (2.20)$$

in which n is an integer, and

$$k^2 = \frac{n^2\pi^2}{L^2} \qquad (2.21)$$

We see, therefore, that the spacing between two permitted states depends on the size of the domain. Because the spacing is inversely proportional to L^2, we obtain, in the limit of large systems, what is called a *continuous spectrum* rather than the discrete spectrum obtained for finite systems.

Often one has to consider a slightly more involved limit in which both the volume, V, of the system and the number of particles, N, are infinite, although their ratio remains constant:

$$N \to \infty, \; V \to \infty, \; \frac{N}{V} = \text{constant} \qquad (2.22)$$

This is the *thermodynamic limit*, which plays an important role in the study of thermodynamic behavior in many-body systems.

The distinction between discrete and continuous spectrums is very important for the description of the time evolution of ρ, the density in phase space. If L has a discrete spectrum, the Liouville equation (2.11') leads to a periodic motion. However, the nature of the motion changes drastically if L has a continuous spectrum.

We shall come back to this in the section on decay of unstable particles in Chapter 3. However, it should be noted here that even a *finite classical system* may have a continuous spectrum in contrast with what happens in quantum mechanics.

Equilibrium Ensembles

As mentioned in Chapter 1, the approach to thermodynamic equilibrium is the evolution toward a final state that acts as an attractor for the initial conditions. It is not difficult to guess what this means in terms of the Gibbs distribution function in phase space. Let us consider an ensemble of which all the members have the same energy E. The Gibbs density ρ is zero except possibly on the energy surface defined by the relation

$$H(p, q) = E \qquad (2.23)$$

Initially, we could consider an arbitrary distribution over this energy surface. This distribution then evolves in time according to the Liouville equation. The simplest view of what thermodynamic equilibrium means is to assume that at thermodynamic equilibrium the distribution ρ would become constant on the energy surface. This was the basic idea that Gibbs had, and he called the corresponding distribution the *microcanonical ensemble* (Gibbs 1902). Gibbs was able to show that this assumption leads to the laws of equilibrium thermodynamics (see also Chapter 4). Besides the microcanonical ensemble, he introduced other ensembles, such as the *canonical ensemble* corresponding to systems in contact with a large energy reservoir at uniform temperature T. This ensemble also leads to the laws of equilibrium thermodynamics and allows a remarkably simple molecular interpretation of such thermodynamic properties as equilibrium entropy. However, such matters will not be dealt with here; instead attention will be focused on the basic question: What kind of conditions have to be imposed on the dynamics of a system to ensure that the distribution function will approach the microcanonical or the canonical ensemble?

Integrable Systems

For most of the nineteenth century the idea of *integrable systems* dominated the development of classical dynamics (see Goldstein, 1950). The

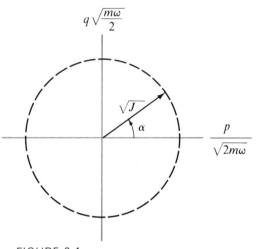

FIGURE 2.4
Transformation from Cartesian coordinates
(p and q) to action and angle variables (J and α,
respectively) for the harmonic oscillator.

idea is easily illustrated by a harmonic oscillator. Instead of the canonical variables q and p, new variables, J and α, are introduced and defined by

$$q = \left(\frac{2J}{m\omega}\right)^{1/2} \sin \alpha$$

$$p = (2m\omega J)^{1/2} \cos \alpha \qquad (2.24)$$

This transformation is quite similar to the transformation from Cartesian to polar coordinates; α is called the angle variable, and J, which is the corresponding momentum, the action variable (see Figure 2.4). With these variables, equation 2.2 takes the simple form

$$H = \omega J \qquad (2.25)$$

We have performed a *canonical transformation* in which one form of the Hamiltonian (2.2) has been changed into another (2.25). What has been gained? In the new form, the energy is no longer divided into kinetic and potential energy. Equation 2.25 gives the total energy directly. We can immediately see the usefulness of such transformations for more complicated problems. As long as we have a potential energy, we cannot really attribute an energy to each of the bodies making up the system, because part of the energy is " between " the various bodies. The canonical

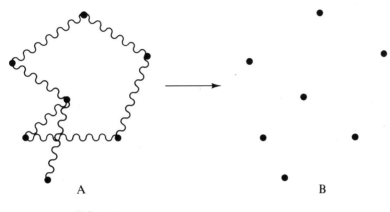

FIGURE 2.5
Elimination of potential energy (represented in part A by
wavy lines) for integrable systems.

transformation gives us a new representation enabling us to speak about
well-defined bodies or particles because the potential energy has been
eliminated. We then obtain a Hamiltonian of the form

$$H = H(J_1, \ldots, J_s) \tag{2.26}$$

which depends only on the action variables. Systems for which we can
transform equation 2.1 into 2.14 and 2.23 into 2.26 through an appropriate change of variables are by definition the integrable systems of
dynamics. For these systems we may therefore "transform away" the
potential energy, as represented schematically in Figure 2.5.

Does the physical world such as the one represented by elementary
particles and their interactions correspond to an integrable system? This
basic question is discussed in Chapter 3.

Another striking feature of the transformation into action and angle
variables is that in equation 2.25 the frequency, ω, of the harmonic oscillator is displayed explicitly in the Hamiltonian (it does not have to be
derived through the integration of the equations of motion). Similarly, for
the general case, we have s frequencies, $\omega_1, \ldots, \omega_s$, each of which is
related to the Hamiltonian by

$$\omega_i = \frac{\partial H}{\partial J_i} \tag{2.27}$$

Coordinates that are by definition the angle variables, $\alpha_1, \ldots, \alpha_s$ correspond to the action variables J_i. Physical quantities are periodic functions of these angle variables.

The form of the Hamiltonian in action variables (equation 2.26) leads to important consequences. The canonical equations are now (see equations 2.4 and 2.27)

$$\frac{d\alpha_i}{dt} = \frac{\partial H}{\partial J_i}, \quad \frac{dJ_i}{dt} = 0, \quad \alpha_i = \omega_i t + \delta_i \tag{2.28}$$

Therefore each action variable is a constant of motion and the angle variables are linear functions of time.

Throughout the nineteenth century, mathematicians and physicists working on problems in classical dynamics looked for integrable systems, because once the transformation into the Hamiltonian form (equation 2.26) has been found, the integration problem (the solution of the equations of motion) becomes trivial. Thus, the scientific community was shocked when Heinrich Bruns first proved (as did Poincaré in more general cases) that the most interesting problems of classical dynamics starting from the three-body problem (e.g., including the sun, the earth, and the moon) do not lead to integrable systems (Poincaré 1889). In other words, we cannot find a canonical transformation that leads to the Hamiltonian form given in equation 2.26; therefore, we cannot find invariants such as the action variables J_i by a canonical transformation. This was in a sense the point at which the development of classical dynamics ended.

Poincaré's basic theorem is discussed later in this chapter in the section titled Dynamical Systems neither Integrable nor Ergodic. For now, it should be noted that, in consideration of the relation of dynamics and thermodynamics, Poincaré's theorem is most fortunate. In general, if physical systems belonged to the class of integrable systems, they could not forget their initial conditions; if the action variables, $J_1, \ldots, J_s$, had prescribed values initially, they would keep them forever, and the distribution function could never become uniform over the microcanonical surface corresponding to a given value E of the energy. Clearly, the final state would drastically depend on the preparation of the system, and concepts such as approach to equilibrium would lose their meaning.

Ergodic Systems

Because of the difficulties in using integrable systems to incorporate the approach to equilibrium, James Clerk Maxwell and Ludwig Boltzmann turned their attention to another type of dynamical system. They introduced what is generally known today as the *ergodic hypothesis*. In the words of Maxwell, "The only assumption which is necessary for a direct proof of the problem of thermodynamic equilibrium is that the system, if left to itself in the actual state of motion, will sooner or later pass through every phase which is consistent with the equation of energy." Mathematicians have pointed out that a trajectory cannot fill "a surface" and that the statement must be altered to indicate that the system will eventually come arbitrarily close to every point of the energy surface, in accord with the quasi-ergodic hypothesis (see Farquhar, 1964).

It is interesting to note that we are dealing with a prototype of dynamical systems, which is just the opposite of the point of view taken in the study of integrable systems. In this prototype, essentially only a single trajectory "covers" the energy surface. Ergodic systems have only one invariant instead of the s invariants, $J_1, J_2, \ldots, J_s$, of integrable systems. If we keep in mind that we are generally interested in many-body systems for which s is of the order of Avogadro's number, $\simeq 6 \times 10^{23}$, the difference is indeed striking.

There is no doubt of the existence of ergodic dynamical systems, even of a very simple type. An example of ergodic time evolution is the motion on a two-dimensional unit square corresponding to the equations

$$\frac{dp}{dt} = \alpha \quad \text{and} \quad \frac{dq}{dt} = 1 \tag{2.29}$$

These equations are easily solved to give, with periodic boundary conditions,

$$\begin{aligned} p(t) &= p_0 + \alpha t \\ q(t) &= q_0 + t \end{aligned} \quad (\text{mod } 1) \tag{2.30}$$

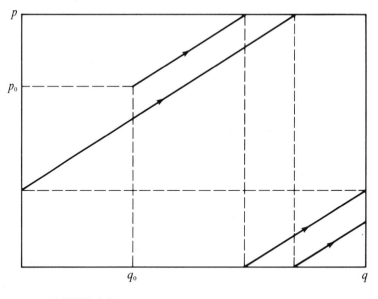

FIGURE 2.6
The phase trajectory given by equation 2.31. For α irrational,
the trajectory is dense on the unit square.

The phase trajectory is therefore

$$p = p_0 + \alpha(q - q_0) \tag{2.31}$$

The basic characteristics of the trajectory depend on the value of α, for which two cases have to be distinguished. If α is a rational number, say $\alpha = m/n$, the trajectory will be periodic and will repeat itself after a period, $T = n$. The system is then *not* ergodic. On the other hand, if α is irrational, then the trajectory will satisfy the condition of the quasi-ergodic hypothesis. It will come arbitrarily near each point of the unit square. It will "fill" the surface of the square (Figure 2.6).

For later reference it is important to note that, in spite of the ergodicity of the motion, each small region of the phase fluid moves without deformation, because a small rectangle $\Delta p\, \Delta q$ preserves not only its size but also its form $(d\Delta p/dt = d\Delta q/dt = 0$ as the consequence of equations 2.29). This is in contrast with other types of motion (see Chapter 7 and Appendix A) in which phase-fluid movement leads to violent disturbances.

In the equations of motion (2.29), α and 1 are two characteristic frequencies (ω_1 and ω_2); one of them relates to p and the other to q. We may write

$$\omega_1 = \alpha, \quad \omega_2 = 1$$

Both are constants, as is the frequency of a harmonic oscillator (see equation 2.25).

When more than one frequency is included in a problem of dynamics, a basic question is the so-called linear independence of the frequencies. If α is rational, we may find numbers m_1 and m_2, both of which do not vanish, such that

$$m_1\omega_1 + m_2\omega_2 = 0 \qquad (2.32)$$

The frequencies are then linearly *dependent*. On the other hand, if α is irrational, equality 2.24 cannot be satisfied with nonvanishing numbers m_1 and m_2. The frequencies are then linearly *independent*.

About 1930, the work of George Birkhoff, John von Neumann, Heinz Hopf, and others gave definite mathematical form to the ergodic problem in classical mechanics. (For references see Farquhar, 1964, and Balescu, 1975.) We have seen that the flow in phase space preserves volume (or "measure"). This still leaves many possibilities open. In an ergodic system, the phase fluid sweeps the whole available phase space on the microcanonical surface, but as we have seen it may do so without altering its shape. But much more complicated types of flow are possible: not only does the phase fluid sweep the entire phase volume, but the initial shape of the element becomes greatly distorted. The initial volume sends out amoebalike arms in all directions in such a way that the distribution, becomes uniform over a long period of time, regardless of its initial configuration. Such systems, called *mixing systems*, were first investigated by Hopf. There is no hope of drawing a simple figure that could correspond to this flow, because two neighboring points, no matter how close together, may diverge. Even if we start with a simply shaped distribution, in time we obtain a "monster," as Benoit Mandelbrot has rightly called objects of this complexity (Mandelbrot 1977). Perhaps a biological analogy can clarify the degree of this complexity: for example, the volume of a lung and the hierarchy of vesicles that it contains.

There are flows with even stronger properties than those of mixing, which have been investigated notably by Kolmogoroff and Ya. Sinai (see Balescu, 1975). Of particular interest are the *K-flows*, whose properties are nearer to those of stochastic systems. In fact, when we go from

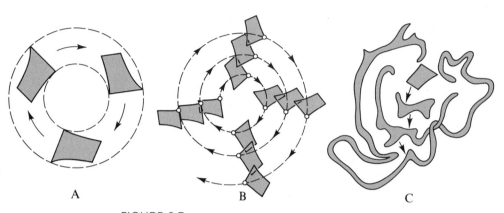

FIGURE 2.7
Various types of flow in phase space:
(A) nonergodic; (B) ergodic but not mixing; (C) mixing.

MODEL OF THE LUNG

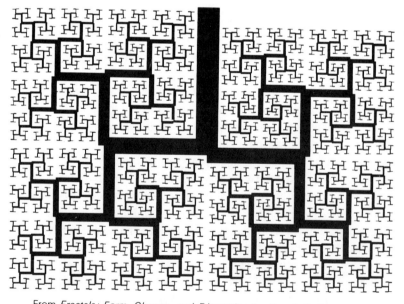

From *Fractals: Form, Chance, and Dimension* by Benoit B. Mandelbrot
(W. H. Freeman and Company). Copyright © 1977 by Benoit B. Mandelbrot.

ergodic flows to mixing flows and then to K-flows, the motion in phase space becomes more and more unpredictable. We are further and further away from the idea of determinism, which was considered the characteristic feature of classical dynamics for such a long time. (An example using the baker transformation is treated in Appendix A.)

With regard to the spectral properties of L, the distinction between these different types of flows is very simple. For example, an ergodic system means that the *only* solution of

$$L\phi = 0 \tag{2.33}$$

is

$$\phi = \phi(H) \tag{2.34}$$

and therefore corresponds to a constant on the microcanonical surface.

By referring to equation 2.13, we can see that equation 2.34 is indeed a solution of equation 2.33, but the characteristic feature of ergodic systems is that it is the *only* one. Similarly (see, e.g., Lebowitz, 1972), mixing implies the stronger property that L has no discrete eigenvalues other than zero. Finally K-flows imply that, in addition to mixing, the multiplicity of solutions (i.e., the number of solutions for a given eigenvalue) is constant.

An unexpected result of ergodic theory is that "unpredictability" or "randomness" of motion is related to such simple properties of the Liouville operator L. In a series of remarkable papers, Sinai (see, e.g., Balescu, 1975) was able to prove that a system of more than two hard spheres in a box was a K-flow (and therefore also mixing and ergodic). Unfortunately, it is not known if this remains true for other (less singular) laws of interaction. Nevertheless, most physicists shared the opinion that this was only a formal difficulty and that the mechanical basis of the approach to equilibrium observed in physical systems had indeed to be found in the theory of ergodic systems.

The view that dynamical systems would be, in general, ergodic was first challenged in a paper by Kolmogoroff (1954). He pointed out that, for large classes of interacting dynamical systems, one could construct periodic orbits confined to a subspace (invariant *tori*) of the ergodic surface. Other investigations also contributed to weakening our belief in the universality of ergodic systems. For example, an important piece of work was the one realized by Enrico Fermi, John Pasta, and Stanislaw Ulam (see Balescu, 1975), who made a numerical investigation of the behavior of a coupled chain of anharmonic oscillators. They expected that this system would reach thermal equilibrium rapidly. Instead, they found periodic oscillations in the energy of the various normal modes. The work of Kolmogoroff was extended by Arnol'd and Moser and has led to the so-called KAM theory. Perhaps the most interesting aspect of this new theory is that, *independently* of ergodicity, dynamical systems may lead to random motion that is somewhat similar to the type of motion occurring in mixing systems or K-flows. Let us consider this important point in more detail.

Dynamical Systems neither Integrable nor Ergodic

To have a clear idea of the behavior of dynamical systems, it is most useful to turn to numerical computations. Work in this direction was pioneered by Michel Henon and Carl Heiles in 1964 (see References) and has since been further developed by many others, such as John Ford and his co-workers (see Balescu, 1975). Generally, the systems used in such computations have two degrees of freedom and the calculations are made for a given value of the energy. Thus, three independent variables remain (because imposing the total energy gives one condition involving the two momenta, p_1, p_2, and the two coordinates, q_1, q_2). A computer program is then worked out to solve the equations of motion and to plot the intersection points of the trajectory with the q_2, p_2-plane. To simplify matters further, one plots only half of these intersections—namely, those at which the trajectory goes "up"; that is, $p_1 > 0$ (see Figure 2.8).

The dynamical behavior of the system can be clearly read in these plots, which had already been used by Poincaré. If the motion is periodic, the intersection is simply a point. If the trajectory is conditionally periodic—that is, if it is restricted to a torus—the successive intersections follow a closed curve in the q_2, p_2-plane. If the trajectory is "random," in the sense that it wanders erratically through the phase space, the intersection point also wanders erratically through the plane. These three possibilities are represented in Figure 2.9.

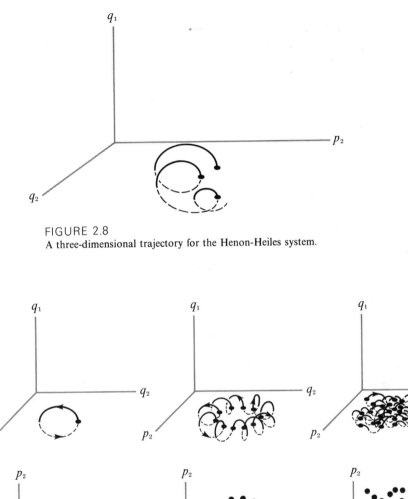

FIGURE 2.8
A three-dimensional trajectory for the Henon-Heiles system.

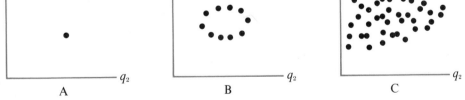

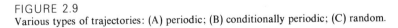

FIGURE 2.9
Various types of trajectories: (A) periodic; (B) conditionally periodic; (C) random.

An interesting observation made by Ford and others is that a dynamical system may, depending on circumstances, change from being conditionally periodic to being "random." To analyze this finding, let us start with the Hamiltonian that is formed by the sum of an unperturbed Hamiltonian, H_0, depending only on canonical momenta and a perturbation depending both on canonical momenta and on canonical coordinates:

$$H = H_0(J_1, J_2) + V(J_1, J_2, \alpha_1, \alpha_2) \tag{2.35}$$

If the perturbation were absent, J_1 and J_2 would be the action variables corresponding to the problem, and we would have two "unperturbed" frequencies related to the Hamiltonian H_0 and given (as in equation 2.27) by

$$\omega_1 = \frac{\partial H_0}{\partial J_1}, \quad \omega_2 = \frac{\partial H_0}{\partial J_2} \tag{2.36}$$

An essential difference between this example and that of the harmonic oscillator is that, in general, H_0 will not be linear in the J's and these two frequencies will be action dependent.

Let us now examine the effect of the perturbation V in the Hamiltonian (equation 2.35). Because this is in general a periodic function in the angle variables, α_1, α_2, we may write it in the general form of a Fourier series. Typically, we may consider a perturbation of the form

$$V = \sum_{n_1, n_2} V_{n_1 n_2}(J_1, J_2) e^{i(n_1 \alpha_1 + n_2 \alpha_2)} \tag{2.37}$$

The interesting point is that the solution of the equation of motion through perturbation theory always includes terms of order:

$$\frac{V_{n_1 n_2}}{n_1 \omega_1 + n_2 \omega_2} \tag{2.38}$$

which correspond to ratios of the potential energy divided by sums of the frequencies for the unperturbed system. This leads to "dangerous" behavior when the Fourier coefficient, $V_{n_1 n_2}$, does not vanish in the presence

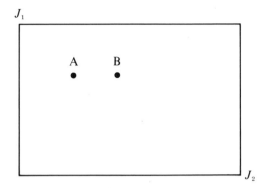

J_1

J_2

FIGURE 2.10
Whittaker's theory (see text for details).

of *resonance* for which

$$n_1\omega_1 + n_2\omega_2 = 0 \qquad (2.39)$$

Expression 2.38 is undefined and anomalous behavior has to be expected.

As shown by the numerical experiments, the occurrence of resonances causes periodic or quasi-periodic behavior to become random behavior (see Figure 2.9). Resonances destroy the simplicity of dynamical motion. They correspond to the transfer of large amounts of energy or momentum from one degree of freedom to the other. In numerical calculations only a finite number—for example, two resonances—are generally considered. But it is important to investigate what would happen if the number of resonances were infinite; that is, if there were resonances in every region of the J_1, J_2-plane, no matter how small. This is the case corresponding to Poincaré's theorem on the nonexistence of integrable systems mentioned earlier. The resonances lead to such an irregular motion that invariants of motion other than the Hamiltonian are no longer analytic functions of the action variables. We shall refer to it as the "Poincaré catastrophe," which will play an important role in the later chapters of this book. It is remarkable how prevalent Poincaré's catastrophe is. It appears in most problems of dynamics starting from the celebrated three-body problem.

A good illustration of the physical meaning of Poincaré's fundamental theorem has been provided by Edmund Whittaker's theory of "adelphic integrals" (1937). Consider a trajectory that starts at some point A in the action space J_1, J_2 of Figure 2.10 and the frequencies, ω_1, ω_2, at this point. Whittaker was able to solve the problem of motion formally for a large class of Hamiltonians in terms of series expansion, but the type of

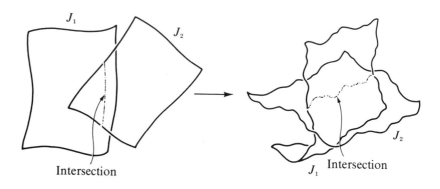

FIGURE 2.11
Trajectory in phase space as the intersection of two invariant surfaces.

series expansion differs crucially according to whether or not the frequencies are rationally independent (or commensurate). Because ω_1, ω_2 are generally continuous functions of the action variables, they will be rationally dependent every time their ratio is a rational number of the form m/n and will be rationally independent if the ratio is not a rational number. Therefore, the type of motion is different for two points A and B even if they are very near one another, because each rational number is embedded in irrationals and vice versa. This is the basic content of the concept of weak stability already mentioned. It is clear that Poincaré's catastrophe may lead to "random" motion. For integrable systems, a trajectory may be viewed as the "intersection" of invariants of motion. For example, in the case of two degrees of freedom a trajectory would correspond to the intersection of the two surfaces, $J_1 = \delta_1$ and $J_2 = \delta_2$, in which δ_1, δ_2 are given constants (see equation 2.26). But whenever we have the Poincaré catastrophe the invariants of motion become nonanalytical, "pathological" functions, as does their intersection (Figure 2.11).

It should be noted that the situation is more complex for nonintegrable systems in which the Poincaré catastrophe arises than it is for ergodic (or mixing) systems. In the first case we know that, as a result of the Kolmogoroff, Moser, and Arnol'd theory, in general *both* periodic motions confined to some part of the available phase space *and* random motions "covering" the whole phase space exist. Both types of motion may have a positive measure. On the contrary, the confined motions of ergodic (or mixing) systems have a measure zero. The consequences of this situation are analyzed in the next section.

Weak Stability

As we have seen, there are at least two types of situations in which dynamical motion introduces random elements. The first corresponds to mixing flow (or flows satisfying stronger conditions, such as K-flows), and the second to what is referred to as the Poincaré catastrophe, in which resonances prevent the "continuation" of the unperturbed invariants of motion when an interaction is initiated. The two situations are quite different: in the first case, the dynamical systems are characterized by a Liouville operator with well-defined spectral properties (such as a continuous spectrum); in the second, it is the decomposition of H (see equation 2.35) into the *two* parts H_0 and V that is essential. In both cases, however, the character of the motion is such that two trajectories, regardless of how close together their starting points are, may diverge greatly in time. This corresponds to what often has been called *instability of motion* and is of obvious importance for the long-term behavior of dynamical systems. To contrast this behavior with the one found in simple systems, let us consider a simple pendulum for which the Hamiltonian is

$$H = \frac{p^2}{2ml^2} - mgl \cos \theta \qquad (2.40)$$

in which the first term is the kinetic energy and the second the potential energy in the gravitational field. The coordinate q is replaced by θ, the angle of deflection.

Such a pendulum can move in two ways: it can either oscillate around its equilibrium position or rotate around its point of suspension. Rotation is possible only when the energy of the pendulum is large enough. The region in which one or the other motion is possible can be represented in phase space, as shown in Figure 2.12. The important point for us is that the neighboring points in phase space corresponding to vibration or rotation belong to the same region. Therefore, even with limited information about the initial state of the system, we can decide if the system will rotate or vibrate.

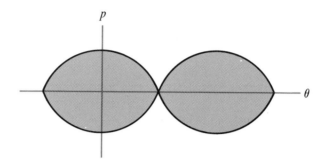

FIGURE 2.12
Phase space for the rotator. Shaded region corresponds to vibration;
outside region corresponds to rotation.

This property is lost for systems in which stability is weak. In such systems, one type of motion may occur in every neighborhood of another type of motion (see Figure 2.13). There is then no point in increasing the precision of our observation. The microstructure of the phase space has become extremely complex. This is the reason why statistical arguments enter into every long-term prediction.

In such situations, statistical ensembles must be considered. We cannot reduce the "mixture" to a "pure" case corresponding to a single trajectory (which would be represented by a δ-function in phase space). Is this difficulty *practical* or *theoretical* in nature? I would support the view that this result has important theoretical and conceptual significance because it forces us to transgress the limits of a purely dynamical description. A similar problem—Is the limitation of the propagation of signals by the velocity of light a *practical* or *theoretical* question?—is answered by the theory of relativity, which shows that our concepts of space and time have to be changed because of this limitation.

There is always the temptation to try to describe the physical world as if we were not part of it. We could then conceive of velocities of propagation of arbitrary, even infinite, speed and the determination of initial conditions with infinite precision. But seeing the world from the outside is not the object of physics. Rather, it is to describe the physical world as it appears to us, who belong to it, through our measurements. In the line of thought inaugurated by the theory of relativity and followed by quantum mechanics, it is a basic objective of theoretical physics to make explicit the general limitations introduced by the measurement processes.

But weak stability is only one step toward the incorporation of time and irreversibility into the formal structure of dynamics. As will be seen,

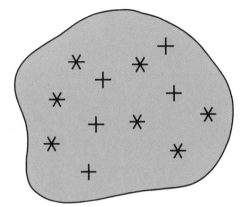

FIGURE 2.13
In systems having weak stability,
one type of motion, ✳, may be found
in the neighborhood of another, +.

the introduction of entropy or, in general, a Lyapounov function greatly
alters this whole formal structure (see Chapters 3 and 7). This is a most
unexpected development. We were prepared to see new theoretical structures arise as a result of discoveries in the field of elementary particles or as
a result of new insights into the evolution of the universe, but that the
concept of thermodynamic irreversibility, which has been with us for
one hundred fifty years, should force us to invent new theoretical structures
is most surprising.

Emphasis should also be placed on the creative role that the problem of
irreversibility has played in the history of classical dynamics, and even
more so in quantum dynamics (see Chapter 3). The challenge of thermodynamics, which led to ergodic theory and to the ensemble theory, has
been the starting point of quite remarkable developments. This produc-
tive dialogue between the physics of being and the physics of becoming is
still going on today, as will be seen in Chapters 7 and 8.

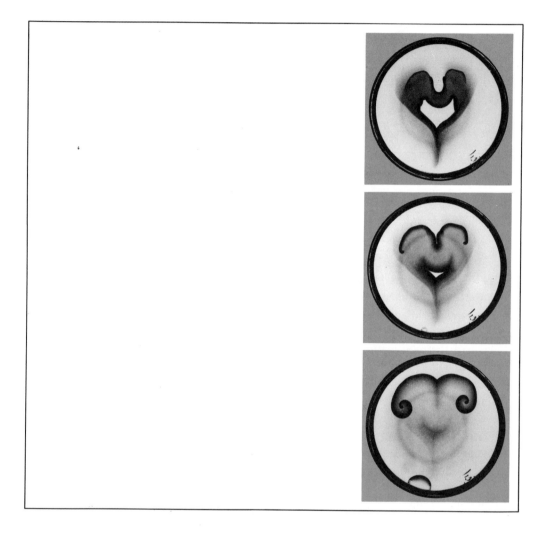

Chapter 3

QUANTUM MECHANICS

Introduction

As demonstrated in Chapter 2, it is only recently that we have begun to
grasp the complexity of dynamical description, even in the framework of
classical dynamics. Still, classical dynamics attempted to represent some
intrinsic reality independent of the mode of description. It was quantum
mechanics that shook the Galilean foundations of physics. It destroyed
the belief that physical description is realistic in a naive sense, that the
language of physics represents the properties of a system independent of
the conditions of experimentation and measurement.

Quantum mechanics has a very interesting history (Jammer 1966;
Mehra 1976, 1979). It started with Planck's attempt to reconcile dynamics
with the second law of thermodynamics. Boltzmann had considered this
problem for interacting particles (which will be discussed in Chapter 7),
whereas Planck thought it would be easier to study the interaction of
matter with radiation. He failed in this purpose, but in his attempt he
discovered the well-known universal constant h, which bears his name.

For a time, quantum theory remained associated with thermodynamics in the theory of black-body radiation and the theory of specific heat. When Arthur Haas presented what may be considered a precursor of Niels Bohr's theory of electronic orbits in 1908 in Vienna as a part of his dissertation, it was refused on the ground that quantum theory had nothing to do with dynamics.

The situation changed drastically when the extraordinary success of the Bohr-Sommerfeld model of the atom made clear the necessity of building a new dynamics in which Planck's constant could be consistently incorporated. This was accomplished by Louis de Broglie, Werner Heisenberg, Max Born, Paul Dirac, and many others.

Because the scope of this book precludes a detailed account of quantum mechanics, the following discussion will focus on the notions necessary for our inquiry: the role of time and irreversibility in physics.

The "classical" quantum theory as formulated in the mid-1920s was inspired by the Hamiltonian theory summarized in Chapter 2. Like this Hamiltonian theory, the quantum theory was immensely successful for such simple systems as the rotator, the harmonic oscillator, or the hydrogen atom. However, as in classical dynamics, problems arise when more complicated systems are considered.

Can quantum mechanics consistently incorporate the concept of elementary particles? Can it describe decay processes? These are the problems to be emphasized here. They are addressed again in Part III of this book in discussing the bridge from being to becoming.

Quantum mechanics is a microscopic theory in the sense that it was introduced with the primary purpose of describing the behavior of atoms and molecules. Thus, it is surprising that it has led to the questioning of the relation between the microworld that we seek to observe and the macroworld to which we ourselves and our measuring devices belong. It can be said that quantum mechanics makes explicit the conflict (which, before its advent, had been implicit) between the dynamical description and the process of measurement (see d'Espagnat, 1976; Jammer, 1974). In classical physics, rigid rods and clocks are often used as models of ideal measurement. They were the main tools used by Einstein in his thought experiments, but there is a supplementary element in measurement, which was emphasized by Bohr. Every measurement is intrinsically irreversible: recording and amplification in measurement are coupled to irreversible events, such as the absorption or emission of light. (See Rosenfeld, 1965, and George, Prigogine, and Rosenfeld, 1973.)

How can dynamics, which treats time as a parameter that has no preferential direction, lead to the element of irreversibility inseparable from measurement? This problem currently attracts a great deal of attention. It is perhaps one of the hottest problems of our time, one in which science and philosophy merge: Can we understand the microscopic world in "isolation"? In fact, we know matter, especially its microscopic properties, only by means of measuring devices, which themselves are macroscopic objects consisting of a large number of atoms or molecules. In a way these devices extend our sense organs. The apparatus can be said to be the mediator between the world that we explore and ourselves.

We shall see that the state of a quantum system is determined by the *wave function*. This wave function satisfies a dynamical equation that is reversible in time, as do the equations of classical dynamics. Therefore, this equation cannot by itself describe the irreversibility of measurement.

The novel aspect of quantum mechanics is that we need both reversibility and irreversibility. In a sense, this was already true in classical physics, in which both types of equations were used: for example, Hamilton's equations of dynamics, which are reversible in time, and Fourier's equation for the temperature evolution, which describes an irreversible process. There, however, the problem could be brushed aside by qualifying the heat equation as a phenomenological equation devoid of any fundamental significance. But how do we brush aside the problem of measurement, which is our very link with the physical world?

Operators and Complementarity

The observation that sharp absorption or emission lines exist has been most important in the formulation of quantum mechanics. The only possible interpretation seems to be that a system like an atom or a molecule *has* discrete energy levels. To reconcile this with the classical ideas, a very important step had to be made. The Hamiltonian, as introduced in Chapter 2, can take a continuous set of values according to the values of its arguments, the coordinates and momenta. Therefore, it seemed necessary to replace the Hamiltonian, H, viewed as a continuous function, with a new object, the Hamiltonian regarded as an operator, and denoted by H_{op}. (For an introduction to quantum mechanics, see Landau and Lifschitz, 1960.)

The concept of operators in connection with classical mechanics was briefly discussed in Chapter 2. However, the situation is quite different in quantum mechanics. In considering trajectories in classical mechanics, we need only the Hamiltonian as a function of coordinates and momenta (see equation 2.4). However, even in the simplest quantum case, such as the interpretation of the properties of the hydrogen atom, we need the *Hamiltonian operator*, because we want to interpret the energy levels as the *eigenvalues* associated with this operator (see equation 2.16). Therefore we must set up and solve the eigenvalue problem

$$H_{op} u_n = E_n u_n \tag{3.1}$$

The numbers $E_1, E_2, \ldots, E_n$ are the energy levels of the system. Of course, we must have rules by which to change from classical variables to quantum operators. One such rule is

$$q \to q_{op}, \quad p \to p_{op} = \frac{h}{i} \frac{\partial}{\partial q} \tag{3.2}$$

which is to say, without going into detail: "Keep the coordinates as they are and replace momenta by derivatives with respect to coordinates."*

In a sense, the transition from functions to operators was forced upon us by spectroscopic experiments that revealed the existence of energy levels. It was a natural step to take, and yet we can only admire people like Max Born, Pascual Jordan, Werner Heisenberg, Erwin Schrödinger and Paul Dirac who dared to make this jump. The introduction of operators radically changes our description of nature. Thus, it is quite appropriate to speak of the "quantum revolution."

To give an example of these new features, the operators that we have to introduce generally do not commute. This has the following consequences: an eigenfunction of an operator is considered to describe the

---

* When there is no possibility of confusion, the subscript "op" will be omitted and H will be used instead of H_{op}.

state of the system in which the physical quantity represented by this operator has a well-defined value (the eigenvalue). Therefore, noncommutativity means, in physical terms, that there can be no state in which, for example, the coordinate q and the momentum p have well-defined values simultaneously. This is the content of the well-known Heisenberg uncertainty relations.

This consequence of quantum mechanics is quite unexpected, as it forces us to give up the naive realism of classical physics. We can measure the momentum and the coordinate of a particle. We cannot say that it *has* well-defined values of coordinate and momenta simultaneously. This conclusion was reached fifty years ago by Heisenberg and Born, among others. It seems as revolutionary today as it did then. In fact, discussions about the meaning of the uncertainty relations have never ceased. Can we not through the introduction of some supplementary "hidden" variables restore physical sanity? Until now this has proved to be difficult, if not impossible, and most physicists have given up such attempts. Although the history of this fascinating subject cannot be related here, it is treated quite well in specialized monographs (see Jammer, 1974).

Niels Bohr formulated the principle of *complementarity* based on the existence of physical quantities represented by noncommuting operators (see Bohr, 1928). I hope that he and my late friend Leon Rosenfeld would not have disapproved of the way in which I would like to define this complementarity: the world is richer than it is possible to express in any single language. Music is not exhausted by its successive stylizations from Bach to Schoenberg. Similarly, we cannot condense into a single description the various aspects of our experience. We must call upon numerous descriptions, irreducible one to the other, but connected to each other by precise rules of translation (technically called transformations).

Scientific work consists of elective exploration rather than a discovery of a given reality; it consists of choosing the problem that must be posed. But rather than anticipate some of the conclusions that are presented in Chapter 9, let us resume the discussion of quantum mechanics.

Quantization Rules

Eigenfunctions play very much the same role as basic vectors in vector algebra. As is known from elementary mathematics, an arbitrary vector,

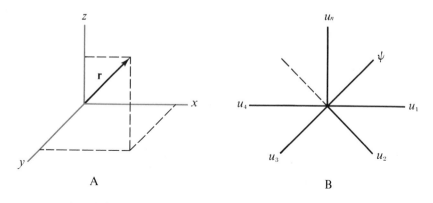

FIGURE 3.1
Decomposition of (A) a vector, **1**, into its components and (B) a wave function, Ψ, into eigenfunctions $u_1, u_2, \ldots, u_n$.

say ℓ, can be decomposed into its components along the set of basic vectors (see Figure 3.1). Similarly, we may represent an arbitrary state Ψ of a quantum mechanical system as a superposition of suitable eigenfunctions:

$$\Psi = \sum c_n u_n \tag{3.3}$$

For reasons that will become apparent in the next section, Ψ is also called the wave function. It is especially convenient to take an orthonormal set of eigenfunctions (the corresponding basic vectors would have length one and be orthogonal to each other):

$$\langle u_i | u_j \rangle = \delta_{ij} \begin{cases} = 1 \text{ if } i = j \\ = 0 \text{ if } i \neq j \end{cases} \tag{3.4}$$

The notation $\langle u_i | u_j \rangle$ indicates the scalar product

$$\langle u_i | u_j \rangle = \int u_i^\dagger u_j \, dx \tag{3.5}$$

in which $u_i^\dagger$ is the complex conjugate to u_i. By multiplying equation 3.3 by $u_m^\dagger$ and using the orthonormality conditions given in equation 3.4, we see immediately that

$$c_m = \langle u_m | \Psi \rangle \tag{3.6}$$

The main difference between the elementary vector space (see Figure 3.1A) and the space used in quantum mechanics (see Figure 3.1B) is in the number of dimensions that are finite in the first case and infinite in the second. In the second case, one speaks of *Hilbert space*, and the functions u_n, or Ψ, are elements (or *vectors*) of this space. Each element may appear in two ways at the left or at the right inside the scalar product (equation 3.5). For this reason, Dirac (1958) introduced an elegant notation. The element u_n may be written either as a *bra* vector

$$\langle u_n |$$

or as a *ket* vector

$$| u_n \rangle$$

The scalar product is then the product of a *bra* with a *ket*:

$$\langle u_n | u_m \rangle$$

This notation allows us to express in a compact way important properties of the Hilbert space. Suppose that the expansion in equation 3.3 is valid for all elements. Using the *bra-ket* notation and equation 3.6, we may then write for an arbitrary element Φ

$$| \Phi \rangle = \sum_n c_n | u_n \rangle = \sum_n | u_n \rangle \langle u_n | \Phi \rangle$$

Because this relation must be true for an arbitrary $| \Phi \rangle$, we obtain the *completeness relation*

$$1 = \sum_n | u_n \rangle \langle u_n | \tag{3.7}$$

which we shall use repeatedly. From this short excursion into the formalism, let us return to physics.

The expansion coefficients c_n, which appear in equation 3.3, have an important physical meaning. If we measure the physical quantity (say, the energy) of which the u_n are the eigenvectors, the probability of finding the eigenvalue corresponding to u_n (say, E_n) is $|c_n|^2$. The function Ψ, which gives the quantum state, is therefore called a *probability amplitude* (its square gives the probability proper). This remarkable physical interpretation of Ψ is due to Born (see Jammer, 1966).

It has already been noted that in quantum mechanics physical quantities are represented by operators. However, these operators cannot be arbitrary. The specific class of operators of interest may be defined by associating each operator A with its *adjoint* $A^\dagger$:

$$\langle v | Au \rangle = \langle A^\dagger v | u \rangle \tag{3.8}$$

A fundamental role is played in quantum mechanics by self-adjoint (or Hermitian) operators:

$$A = A^\dagger \tag{3.9}$$

Their importance stems from the fact that eigenvalues of self-adjoint or Hermitian operators are real. Moreover, a Hermitian operator leads to an orthonormal set of eigenfunctions satisfying condition 3.1. It is often stated that "*observables*" *are represented in quantum mechanics by Hermitian operators*. Are all observables Hermitian operators? This is a complicated question, which is dealt with in Chapter 8.

In addition to Hermitian operators we need a second class of operators that are associated with changes in coordinates. It is known from elementary geometry that coordinate changes do not alter the value of a scalar product. Therefore let us consider operator A such that it leaves the scalar product (equation 3.5) invariant. This implies

$$\langle Au | Av \rangle = \langle u | v \rangle \tag{3.10}$$

and, as a consequence, using equation 3.8,

$$A^\dagger A = 1 \tag{3.11}$$

By definition, operators satisfying equation 3.11 are called *unitary* operators. The inverse of the operator A is A^{-1}, such that

$$AA^{-1} = A^{-1}A = 1 \tag{3.11'}$$

We therefore see that a unitary operator is characterized by the property that its inverse is equal to its adjoint:

$$A^{-1} = A^\dagger \qquad (3.12)$$

As in elementary geometry, a similitude transformation must often be performed on operators. A similitude S leads from A to $\tilde{A}$ through the relation

$$\tilde{A} = S^{-1}AS \qquad (3.13)$$

An interesting property is that such similitudes leave invariant all algebraic properties; for example, if

$$C = AB, \quad \text{then} \quad \tilde{C} = \tilde{A}\tilde{B} \qquad (3.14)$$

because, using equation 3.11',

$$\tilde{C} = S^{-1}ABS = (S^{-1}AS)(S^{-1}BS)$$

The similitude (in equation 3.13) may be considered a mere change in coordinates if S is a unitary operator. We are now ready to formulate the problem of quantization as one of finding a suitable coordinate system in which the Hamiltonian takes a simple, diagonal form. This is the Born-Heisenberg-Jordan quantization rule (see Dirac, 1958).

We start with the Hamiltonian containing, as in equation 2.1, a kinetic (or *unperturbed*) contribution, H_0, plus a potential energy (or *perturbation*), V. We may then look for a similitude

$$\tilde{H} = S^{-1}HS \qquad (3.15)$$

in terms of a unitary operator S, which transforms the initial Hamiltonian into a diagonal one. This is equivalent to the solution of the eigenvalue problem in equation 3.1. Indeed, we may represent H as a matrix, and equation 3.1 shows that, in the representation using its eigenfunctions, H is represented by a diagonal matrix:

$$\langle u_i | H u_j \rangle = E_j \langle u_i | u_j \rangle = E_j \delta_{ij} \qquad (3.16)$$

The analogy with the transformation problem of classical mechanics considered in Chapter 2 in the section on integrable systems is striking.

The Born-Heisenberg-Jordan quantization rules will be returned to in Chapter 8 in a discussion of how the systems that display irreversible processes can be quantized. For now, let it suffice to note that, as in classical transformation theory, two possible descriptions of a physical system have been applied to integrable systems. Indeed, diagonalization of the Hamiltonian is quite similar to the classical transformation of the Hamiltonian to action variables (equation 2.26).

This point can be simply illustrated by a harmonic solid, which corresponds to interacting neighboring atoms or molecules whose relative displacement is so small that it can be described in terms of a potential energy that is quadratic in the displacement, such as that in the harmonic oscillator (equation 2.2). We may describe this system in two ways. The first corresponds to the interaction between neighboring particles in the solid, in which case we have to consider both the kinetic and the potential energy (refer to Figure 2.5A). The second way requires, as in the section on integrable systems in Chapter 2, a canonical transformation to eliminate the potential energy. We may then consider the solid to be a superposition of independent oscillators and calculate the energy levels of each oscillator (refer to Figure 2.5B). Again, we have a choice of descriptions: one in which entities are not well defined (because part of the energy of the solid is "between" the particles) and the other in which they are independent, the "normal modes" of the solid. We return once again to the question, Does our physical world belong to one of these two highly idealized descriptions, or is a third one necessary? This question will be further dealt with later in this chapter in the section on ensemble theory in quantum mechanics.

Time Change in Quantum Mechanics

In the preceding section, the concept of the state of a quantum system as described by some state vector Ψ was introduced. We now need an equation that will describe its time variation. This equation must play the same role in quantum mechanics as do Hamilton's equations (2.4) in classical mechanics. The analogy that guided Schrödinger in formulating this new equation was with classical optics—the eigenvalues corresponding to the characteristic frequencies associated with wave phenomena. Schrödinger's equation is a *wave* equation involving the basic dynamical quantity, the Hamiltonian. Its explicit form is

$$i\hbar \frac{\partial \Psi}{\partial t} = H_{\text{op}} \Psi \qquad (3.17)$$

in which i is the symbol $\sqrt{-1}$ and $\hbar$ is Planck's constant divided by 2π (we shall often take $\hbar$ equal to one, to avoid excessive notation). Note that this equation is not derived in quantum mechanics, but assumed. It can be validated only by comparison with experiment.

The Schrödinger equation is a *partial* differential equation (in that derivatives with respect to coordinates appear in H_{op} (see the next section) in contrast with Hamilton's equations (2.4). But they do have an element in common: both Hamilton's equations and Schrödinger's equation are of first order in time. Once Ψ is known at some arbitrary time t_0 (together with suitable boundary conditions such as $\Psi \to 0$ at infinite distances), we may calculate Ψ for arbitrary times both in the future or in the past. In this sense we recover the deterministic view of classical mechanics, but it now applies to the wave function and not to the trajectory, as in classical mechanics.

The discussion in Chapter 2 on the Liouville equation can be applied directly here. It is true that Ψ represents a probability amplitude (as ρ in equation 2.12 represents a probability), but its time evolution has a strictly dynamical character. As in the Liouville equation, there is no simple condition here with a probabilistic process such as Brownian motion.

The time evolution is determined by the Hamiltonian. Therefore, in quantum mechanics, the Hamiltonian (more precisely the Hamiltonian operator) plays a *dual* role. On the one hand, it determines the energy levels through equation 3.1. On the other, it determines the time evolution of the system.

It is also important to notice that the Schrödinger equation is linear. If at a given moment t we have

$$\Psi(t) = a_1 \Psi_1(t) + a_2 \Psi_2(t) \qquad (3.18)$$

then at another arbitrary time t', earlier or later than t, we also have

$$\Psi(t') = a_1\Psi_1(t') + a_2\Psi_2(t') \tag{3.19}$$

We have seen that Ψ determines the probability of the outcome of experiments and may be appropriately called a probability amplitude. It is also called the *wave function*, because equation 3.17 has a strong formal similarity with the wave equations of classical physics.

It is easy to give the formal solution of the Schrödinger equation (3.17):

$$\Psi(t) = e^{-iHt}\Psi\,(t=0) \tag{3.20}$$

This may be verified by taking its derivative.

This form is quite similar to equation 2.12', except that the Liouville operator L is replaced by the Hamiltonian H. Note that e^{-iHt} (or e^{-iLt}) is a unitary operator, in agreement with equation 3.12:

$$(e^{-iHt})^\dagger = e^{iHt} = (e^{-iHt})^{-1}$$

This results from the fact that H is Hermitian. Therefore, in both classical and quantum mechanics, the time evolution is given in terms of a unitary transformation. Time evolution corresponds merely to a change of coordinates!

If we use the expansion (in equation 3.3) of Ψ in terms of the eigenfunctions of the Hamiltonian, we obtain from equation 3.20 the explicit relation

$$\Psi(t) = \sum_k e^{-iE_k t}c_k u_k \tag{3.21}$$

According to our rule, the probability of finding the system in the state u_k will be given by

$$\left|e^{-iE_k t}c_k\right|^2 = |c_k|^2 \tag{3.22}$$

The important point is that this probability is time independent. In the representation in which the energy is diagonal, nothing really "happens." The wave function simply "rotates" in the Hilbert space, and the probabilities are constant in time.

Quantum mechanics may be applied to systems formed by many particles. Here the concept of indistinguishability plays a very important role. Consider, for example, a collection of N electrons. Ψ will now depend on all the N electrons. A permutation of, say, electrons 1 and 2 should not change the physical situation. Therefore, we have to require (remember that Ψ is a probability amplitude and that probabilities are given by $|\Psi|^2$)

$$|\Psi(1, 2)|^2 = |\Psi(2, 1)|^2 \qquad (3.23)$$

We may satisfy this condition in two ways. Either

$$\Psi(1, 2) = +\Psi(2, 1) \qquad (3.24)$$

or

$$\Psi(1, 2) = -\Psi(2, 1) \qquad (3.24')$$

These two ways correspond to the two basic *quantum statistics*: the Bose statistics, when the wave function does not change under the permutation of the two particles, and the Fermi statistics, when it does. The type of statistics seems to be quite a fundamental property of matter, because all known elementary particles obey one or the other. Protons, electrons, and so forth, are *fermions*; photons and some unstable particles, like mesons, are *bosons*. One of the great achievements of quantum mechanics is the discovery of this distinction between fermions and bosons, which shows up at all levels of the structure of matter. The behavior of metals, for example, could not be understood without Fermi statistics, as applied to electrons, and the behavior of liquid helium is a beautiful illustration of Bose statistics. The problem of Bose or Fermi statistics in connection with the decay of quantum states is discussed in the next section.

Ensemble Theory in Quantum Mechanics

Using the formalism of quantum mechanics, we can calculate the average value $\langle A \rangle$ of some dynamic quantity, A, whose eigenvalues are $a_1, a_2, \ldots$. By definition, an average value is the sum of all values, $a_1, a_2, \ldots$, that the variable can take, each being multiplied by the corresponding probability. We therefore obtain, using equation 3.6,

$$\langle A \rangle = \sum_n a_n |c_n|^2 = \sum_n \langle \Psi | u_n \rangle \langle u_n | \Psi \rangle \tag{3.25}$$

Using the definition of the eigenfunctions u_n,

$$A u_n = a_n u_n$$

This can also be written as

$$\langle A \rangle = \langle \Psi | A\Psi \rangle \tag{3.26}$$

The important point is that the average value $\langle A \rangle$ is *quadratic* in the probability amplitude. This is in contrast with equation 2.14, which is *linear* in the Gibbs distribution function ρ. Note also that, in a sense, even a system characterized by a well-defined wave function Ψ already corresponds to an ensemble.

Indeed if we expand Ψ, for example in terms of the eigenfunctions of the Hamiltonian (see equation 3.3), and measure the energy, we may find the eigenvalues E_1, E_2, ..., each with the probability $|c_1|^2$, $|c_2|^2$, ... This seems to be an unavoidable consequence of Born's statistical interpretation of quantum mechanics. As a result, quantum mechanics can only make predictions about "repeated" experiments. In this sense, the situation is similar to that of a classical ensemble of dynamical systems described by a Gibbs ensemble.

Yet, in quantum mechanics there is also a clearcut difference between *pure cases* and *mixtures* (see the section titled Hamiltonian Equations of Motion and Ensemble Theory in Chapter 2). To formulate this difference, it is useful to introduce the quantum analogue of the Gibbs distribution function ρ. To do so, we must first introduce a set of complete orthonormal functions n such that, as in equations 3.4 and 3.7,

$$\langle n|m \rangle = \delta_{nm}, \quad \sum |n\rangle\langle n| = 1 \tag{3.27}$$

We then expand Ψ in terms of the functions n and use equation 3.6. We obtain

$$\langle A \rangle = \langle \Psi | A\Psi \rangle = \sum_n \langle \Psi | n \rangle \langle n | A\Psi \rangle$$

$$= \sum_n \langle n | A\Psi \rangle \langle \Psi | n \rangle \tag{3.28}$$

In classical mechanics, the averaging operation includes the integration over phase space (see equation 2.14). We now introduce the *trace operation*, which plays a similar role in quantum mechanics,

$$\text{tr } O = \sum_n \langle n|On \rangle \tag{3.29}$$

and the density operator ρ, defined by

$$\rho = |\Psi\rangle\langle\Psi| \tag{3.30}$$

Again, this definition uses Dirac's "bracket" notation (see equation 3.7). Operators act on elements of the Hilbert space. For example ρ acting on $|\Phi\rangle$ will be given according to definition 3.30 by

$$\rho|\Phi\rangle = |\Psi\rangle\langle\Psi|\Phi\rangle = \langle\Psi|\Phi\rangle|\Psi\rangle$$

The reason for introducing definition 3.30 is that we may now obtain for the average $\langle A \rangle$ as given by equation 3.28 the compact expression

$$\langle A \rangle = \text{tr}(A\Psi)\langle\Psi)$$

$$= \text{tr } A\rho \tag{3.31}$$

which exactly corresponds to the classical form (2.14), the integration over phase space being replaced by the trace operator.

Alternatively expression 3.31 can be written as

$$\langle A \rangle = \sum_{nn'} \langle n|A|n'\rangle\langle n'|\rho|n\rangle \tag{3.31'}$$

in which we have used the notation

$$\langle n|A|n'\rangle \equiv \langle n|An'\rangle$$

If the observable A is diagonal (i.e., $A|n\rangle = a_n|n\rangle$), expression 3.31' reduces simply to

$$\langle A \rangle = \sum_n \langle n|A|n\rangle\langle n|\rho|n\rangle$$

$$= \sum_n a_n\langle n|\rho|n\rangle \tag{3.31''}$$

Therefore the diagonal elements of ρ may be viewed as the *probabilities* of finding the value a_n of the observable. Note that trace of ρ is unity, as we have (see equations 3.27 and 3.30)

$$\text{tr } \rho = \sum_n \langle n | \Psi \rangle \langle \Psi | n \rangle = \sum_n \langle \Psi | n \rangle \langle n | \Psi \rangle$$

$$= \langle \Psi | \Psi \rangle = 1 \qquad (3.31''')$$

This is the quantum mechanical analogue of equation 2.9.

As in classical mechanics, the interest of the ensemble approach is that we can consider more general situations, for example, corresponding to a weighted superposition of various wave functions. Then equation 3.30 becomes

$$\rho = \sum p_k | \Psi_k \rangle \langle \Psi_k | \qquad (3.32)$$

with

$$0 \leqslant p_k \leqslant 1, \ \sum p_k = 1 \qquad (3.33)$$

in which p_k represents the weights corresponding to the various wave functions Ψ_k, which make up the ensemble.

The form of the density operator ρ permits us to make a clearcut distinction between pure cases corresponding to a simple wave function and mixtures. In the first case, ρ is represented by equation 3.30; in the second, by equation 3.32. This leads to a simple formal distinction. For pure cases,

$$\rho^2 = | \Psi \rangle \langle \Psi, \Psi \rangle \langle \Psi | = | \Psi \rangle \langle \Psi | = \rho$$

and ρ is then an idempotent operator. This is not so for mixtures.

The distinction between pure cases and mixtures is necessary to formulate the measurement problem, as will be seen in a later section titled The Measurement Problem.

Schrödinger and Heisenberg Representations

Once we know the time variation (equation 3.20) of the wave function through the solution of the Schrödinger equation, we immediately obtain (from equation 3.30) the time variation of the density ρ:

$$\rho(t) = e^{-itH}\rho(0)e^{itH} \qquad (3.34)$$

By taking the derivative, this leads to

$$i\frac{\partial\rho}{\partial t} = H\rho - \rho H \qquad (3.35)$$

This equation is valid both for pure cases and for mixtures. We obtain exactly the same type of formula that we derived in classical mechanics (formula 2.11). The only difference is that instead of the Poisson bracket we now have the *commutator* of H with ρ.

To emphasize the similarity between these two situations, we shall write the evolution equation (3.35) and its formal solution again in the form

$$i\frac{\partial\rho}{\partial t} = L\rho, \qquad \rho(t) = e^{-iLt}\rho(0) \qquad (3.36)$$

including the Liouville operator, which now has a new meaning. This will permit us to treat both classical and quantum systems by the same methods in Chapter 7.

Let us have another look at the average value of a mechanical quantity and its time variation. We have, using equations 3.31 and 3.34,

$$\langle A(t)\rangle = \text{tr } A\rho(t) = \text{tr } Ae^{-iHt}\rho e^{iHt}$$

$$= \text{tr}(e^{iHt}Ae^{-iHt})\rho$$

$$= \text{tr } A(t)\rho \qquad (3.37)$$

because the definition of the trace operator (equation 3.29) implies that (see expression 3.31')

$$\text{tr } AB = \text{tr } BA \qquad (3.38)$$

Although operators generally do *not* commute (see the section on operators and complementarity earlier in this chapter), they do so when implied in the trace operation. We have also written ρ instead of $\rho(t = 0)$. We may therefore obtain the average value $\langle A(t)\rangle$ in two equivalent ways. In the first, the density changes in time and A remains constant, whereas, in the second, we consider that the density remains constant but the mechanical quantity A changes according to equation 3.37:

$$A(t) = e^{iHt} A e^{-iHt} \tag{3.39}$$

This second description is called the *Heisenberg representation*. It differs from the *Schrödinger representation* in that, instead of the mechanical quantities like A being considered time-independent, the wave function Ψ or ρ is time-independent. By taking the derivative with respect to time, equation 3.39 leads to (see equations 3.35 and 3.36)

$$i\frac{\partial A}{\partial t} = AH - HA$$

$$= -LA \tag{3.40}$$

Note that it is of the same form as the Liouville equation (3.36), except that L is replaced by $-L$. This will be used in Chapter 7.

A similar distinction exists in classical dynamics. Equation 2.5 corresponds to the Heisenberg equation and equation 2.11 to the Schrödinger equation. These two equations differ by the sign of the Poisson bracket operator L as defined in equation 2.13.

Equilibrium Ensembles

The concept of equilibrium ensembles that was introduced for classical systems in Chapter 2 may be easily extended to quantum systems. Yet, there are interesting differences between classical and quantum dynamical systems. For example, quantum ergodic systems can be shown to imply that the systems are *not degenerate* (to each eigenvalue of the energy corresponds a single eigenfunction). This result, which was established by von Neumann (see Farquhar, 1964), very much limits the interest of the ergodic approach because most quantum systems of interest are degenerate. For example, a given energy may be partitioned in many ways between possible excitations in a many-particle system. For this reason, a number of physicists starting with von Neumann himself have tried to define *macro-observables* that would give an *approximate* description of dynamics and include the approach to equilibrium. Once again, we encounter the idea that approach to equilibrium and more generally the concept of irreversibility correspond to an *approximation* of dynamics. It will be seen in Chapter 7 that we can consider this problem quite differently: irreversibility corresponds indeed to an extension of dynamics, possible when supplementary conditions (such as weak stability in classical dynamics) are satisfied.

The Measurement Problem

Many conceptual problems refer to the very formulation of quantum mechanics. For example: Is the departure from classical causality really unavoidable? Can we not introduce supplementary "hidden" variables so as to make the formalism of quantum mechanics more similar to that of classical mechanics? These questions are beautifully reviewed in a monograph by Bernard d'Espagnat (1976). In spite of the effort expended in attempting to solve such problems, no marked success has been achieved until now. Our attitude will be different: we accept the quantum mechanical formalism but we ask how far we can extend it without marked modifications.

This question arises when the measurement problem mentioned early in this chapter is considered. Suppose that we start with a wave function Ψ and the corresponding density ρ as given by equation 3.30:

$$\Psi = \sum c_n u_n$$

$$\rho = |\Psi\rangle\langle\Psi| = \sum_{nm} c_n c_m^\dagger |u_n\rangle\langle u_m| \tag{3.41}$$

By measuring a dynamical quantity, say the energy, of which the u_n are the eigenfunctions, we obtain various eigenvalues $E_1, E_2, \ldots$, with probabilities $|c_n|^2$. But once we have obtained a given eigenvalue, say E_i, we know that the system is necessarily in the state u_i. At the end of the measurement we have a *mixture*:

$$\Psi \to \begin{matrix} u_1 \\ u_2 \\ \vdots \\ u_k \\ \vdots \end{matrix}$$

with probabilities $|c_1|^2$, $|c_2|^2$, ..., $|c_k|^2$, ... In accordance with equation 3.32, the corresponding density ρ is now

$$\rho = \sum_n |c_n|^2 |u_n\rangle\langle u_n| \tag{3.42}$$

which is quite different from equation 3.41.

The transformation from equation 3.41 to equation 3.42, often called the *reduction of the wave packet*, does not belong to the type of unitary transformations (equation 3.20) described by the solution of the Schrödinger equation. Von Neumann (1955) has expressed this difference in a most elegant way by showing that we may define an "entropy" that increases when we go from a pure state to a mixture. In this way the problem of irreversibility now appears at the very heart of physics.

But how is this problem possible? We have seen that Schrödinger's equation is linear (see equation 3.18). A pure state should therefore remain a pure state. If indeed the "fundamental level" of description is the Schrödinger equation, there is no easy way out. Many suggestions are given in d'Espagnat's book (1976), none quite convincing.

The solution proposed by von Neumann himself (1955) and advocated by others, including Eugene Wigner, is that we have to leave the field of physics and invoke the active role of the observer. This is in line with the general philosophy already mentioned that irreversibility is *not* in nature, but in us. In the present case, it is the *perceiving subject* engaged in an act of observation who decides that a transition between a pure state and a mixture occurs. It is easy to criticize this point of view, but again how do we introduce irreversibility into a "reversible" world?

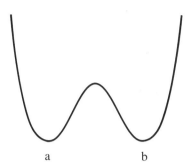

FIGURE 3.2
A symmetric potential.

Others go even further; they claim that there is no reduction of the wave packet at the price that our universe is continuously splitting into a stupendous number of branches as the result of measurementlike interactions! Although such extreme views will not be discussed here, it should be noted that the very existence of such concepts is proof that physicists are not positivists. They are not satisfied with giving rules that simply "work"!

We shall return to this problem in Chapter 8. Let us only note here that the distinction between pure states and mixtures, which is formally very clear in quantum mechanics, can in fact lie beyond any finite accuracy of measurement. For example, consider a symmetric potential with two minima such as that represented in Figure 3.2.

Suppose that $|u_1\rangle$ corresponds to a wave function centered in region "a," and $|u_2\rangle$ to one centered in region "b." The distinction between pure states and mixtures differs by terms involving the product $|u_1\rangle\langle u_2|$. However, this product can be extremely small when the potential barrier takes macroscopic dimensions. In other words, wave functions may become "unobservables," somewhat like the trajectories in problems involving weak stability considered in Chapter 2. This remark will play an important role in the theory of quantum irreversible processes presented in Chapters 7 through 9.

Decay of Unstable Particles

Before discussing the decay of unstable particles, a distinction between "small" and "large" systems should be made clear. The transition from a discrete to a continuous spectrum was treated in Chapter 2 in the section on operators. A general theorem in quantum mechanics states that quantum mechanical systems confined to a *finite* volume have a discrete spectrum. To obtain a continuous spectrum, we must therefore go to the limit of an infinite system. This is in contrast with classical systems in which, as noted in Chapter 2, the Liouville operator may already have a continuous spectrum for *finite* systems. The difference comes from the fact that the classical Liouville operator acts on the phase space, which involves velocities (or momenta) that are always continuous variables, whereas the Hamiltonian operator acts on the coordinate space (or on the momentum space but not on both; see equations 3.1 and 3.2).

A discrete spectrum for H means periodic motion. This is no longer so when the spectrum becomes continuous. Let us therefore see how the transition to a continuous spectrum changes the time evolution. Instead of the sum in equation 3.21, an integral must now be considered. Using the eigenvalue of the energy as the independent variable, we may write this integral in the form

$$\Psi(t) = \int_0^\infty d\varepsilon \ e^{-i\varepsilon t} f(\varepsilon) \tag{3.43}$$

An important point is that this integral has to be taken from a finite value (in this case, a finite value equal to zero) to infinity. Indeed, if the Hamiltonian could take arbitrarily large negative values, the system would be unstable; therefore some lower limit must exist.

Instead of the periodic variation represented by equation 3.21, we now obtain a Fourier *integral*, which may represent a much larger type of variation in time. In principle, this is welcome. We may, for example, apply this formula to the decay of an unstable particle or to the deactivation of an excited atomic level. Then, by introducing appropriate initial conditions, we would like to find an exponential decay for the probability

$$|\Psi(t)|^2 \sim e^{-t/\tau} \tag{3.44}$$

in which τ is the lifetime. This is nearly, but not exactly, so. In fact, exponential formula 3.44 can never be exact. As a consequence of a celebrated theorem, the Paley-Wiener theorem (1934), a Fourier integral of the form 3.43, in which the integration is taken from a finite value to infinity, always decays more slowly than an exponential in the limit of long times. In addition, equation 3.43 leads to short time deviations from the exponential law.

It is true that numerous theoretical investigations have shown that the deviations from the exponential are too small to be measured at present. It is important that experimental and theoretical investigations be continued. The fact that deviations from the exponential law of decay exist leads to serious questions about the meaning of indistinguishability. Suppose that we prepared a beam of unstable particles, say mesons, and let it decay and that later we prepared another group of mesons. Strictly speaking, these two groups of mesons, prepared at two different times, would have different decay laws, and we could distinguish between the two groups just as we can distinguish between old and young women. This seems somewhat strange. If we have to choose, I believe we should keep indistinguishability as a basic principle.

Of course, if we restricted the concept of elementary particles to stable particles—as, for example, Wigner has suggested on many occasions—then this question would not arise. But it seems difficult to restrict the existing schemes for elementary particles to stable ones. It seems fair to say that the scientific community feels more and more that some generalization of quantum mechanics is necessary to incorporate unstable particles. In fact, the difficulty is even greater. We would like to associate elementary particles with well-defined properties *in spite of their interactions*. To take a concrete case, consider the interaction of matter with light, of electrons with photons. Suppose that we could diagonalize the corresponding Hamiltonian. We would obtain some "units" similar to the normal modes of a solid, which by definition no longer interact. Certainly these units cannot be the physical electrons or photons that we see around us. These objects interact and it is precisely because of this interaction that we can study them. But how to incorporate interacting but well-defined objects into a Hamiltonian description? As mentioned before, in the representation in which the Hamiltonian is diagonal the objects are well defined, but there are no interactions; in the other representations the objects are not well defined.

One feels that a way out must be a closer look at what we really have to eliminate and what to keep by a suitable transformation. As will be seen in Chapter 8, this problem is closely related to the basic distinction between reversible and irreversible processes.

Is Quantum Mechanics Complete?

In light of the discussions presented, I believe that the answer to this question can be safely formulated as "no." Quantum mechanics was directly inspired by the situation in atomic spectroscopy. The frequency of "rotation" of an electron around the nucleus is of the order of 10^{-16} second, the typical lifetime 10^{-9} second. Therefore, an excited electron rotates 10,000,000 times before it falls down to the ground state. As it was well understood by Bohr and Heisenberg, it was this fortunate circumstance that made quantum mechanics so successful. But today we can no longer be satisfied with approximations that treat the nonperiodic part of the time evolution as a small, insignificant perturbation effect. Here again, as in the problem of measurement, we are confronted with the concept of irreversibility. With his remarkable physical insight, Einstein noticed (1917) that quantization in the form used at that time (i.e., in the Bohr-Sommerfeld theory) was valid only for quasi-periodic motions (described in classical mechanics by integrable systems). Certainly fundamental progress has been realized since. Yet the problem remains.

We are faced with the very meaning of idealizations in physics. Should we consider quantum mechanics of systems in a *finite volume* (and therefore with a discrete energy spectrum) to be the basic form of quantum mechanics? Then problems such as decay, lifetimes, and so forth must be considered to be related to supplementary "approximations" involving the limit to infinite systems to obtain a continuous spectrum. Or, on the contrary, should we argue that nobody has ever seen an atom that would not decay when brought into an excited level? The physical "reality" then corresponds to systems with continuous spectra, whereas standard quantum mechanics appears only as a useful idealization, as a simplified limiting case. This is much more in line with the view that elementary particles are expressions of basic fields (such as photons with respect to the electromagnetic field) and fields are in essence not local because they extend over macroscopic regions of space and time.

Finally, it is interesting to note that quantum mechanics has introduced statistical features into the basic description of physics. This is most clearly expressed in terms of the Heisenberg uncertainty relations. It is important to note that no similar uncertainty relation exists for time and energy (i.e., the Hamiltonian operator). As the result of Schrödinger's equation, which relates the time change to H_{op}, such an uncertainty relation could be understood as a complementarity between *time* and *change*, between being and becoming. But time is just a number (not an operator) in quantum mechanics, as it is in classical mechanics.

We shall see that there are circumstances—implying the limit to continuous spectrum—in which such a supplementary uncertainty relation may be established between the Liouville operator and time even in classical mechanics. When this is so, time acquires a new supplementary meaning—it becomes associated with an operator. Before we take up this fascinating problem again, let us consider the "complementary" part of physics; that is, the physics of *becoming*.

Part II

THE PHYSICS
OF BECOMING

EMERGENCE OF WAVE STRUCTURES IN FIELDS OF AMOEBAE
OF THE CELLULAR SLIME MOLD *Dictyostelium discoideum.*

When mature slime-mold amoebae have exhausted their food resources
and become starved, they secrete cyclic AMP, an attractant that induces
their aggregation. The attractant is secreted in brief pulses, initially by only
a few amoebae that then become the centers to which other amoebae
are attracted. The frequency of these pulses is one every five minutes at
first, increasing to one every two minutes as aggregation proceeds. The
nine frames shown here were taken at ten-minute intervals. The initial
signals, which decay within a few seconds of their creation, are passed
on to amoebae nearby, and they in turn pass them on to amoebae
further away, and so forth. The pulses are relayed outward from each
center at about one millimeter every three minutes. Besides relaying the
pulses, the amoebae respond to these signals by moving a short distance
toward a signalling center each time a pulse reaches them. This
discontinuous movement has been visualized in these photographs by the
use of a finely adjusted dark-field optical system that reveals the moving
amoebae (which are elongated) as bright bands and the stationary
amoebae (which are rounded up) as dark bands. Waves can be either
concentric, with a period controlled by pace-maker amoebae, or spiral,
with periodicity governed by the refractory interval—that is, the length of
time following a response during which the amoebae will not respond
to further stimulation. The larger circles shown are approximately ten
millimeters in diameter. [Unpublished photographs by P. C. Newell,
F. M. Ross, and F. C. Caddick. Further details of the signalling system
can be found in "Aggregation and Cell Surface Receptors in Cellular Slime
Molds" by P. C. Newell. In *Microbial Interactions, Receptors and
Recognition*, Series B, J. L. Reissig, ed. (Chapman & Hall, 1977),
pp. 1—57.]

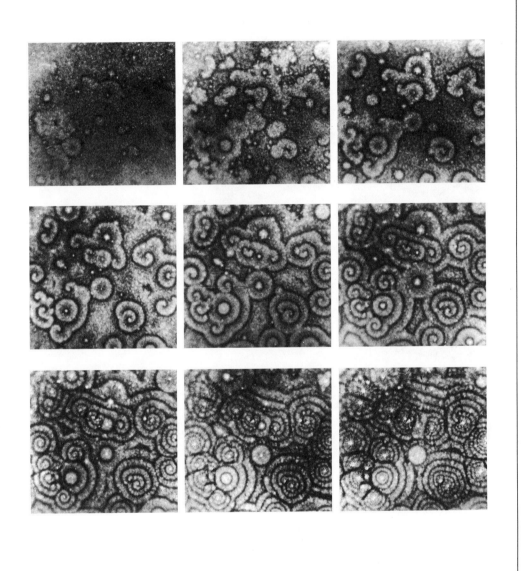

Chapter 4

THERMODYNAMICS

Entropy and Boltzmann's Order Principle

Chapters 2 and 3 of this book dealt with the physics of time corresponding to *reversible* phenomena, because both the Hamilton and the Schrödinger equations are invariant with respect to the substitution $t \rightarrow -t$. Such situations correspond to what I have called the *physics of being*. We now turn to the *physics of becoming* and, specifically, to irreversible processes as described by the second law of thermodynamics. In this chapter and in the two that follow, the point of view is strictly phenomenological. What the relation with dynamics may be will not be investigated; however, methods will be outlined that successfully describe unidirectional time phenomena over a wide range, from simple irreversible processes such as heat conduction to complicated processes involving self-organization.

Since its formulation, the second law of thermodynamics has emphasized the unique role of irreversible processes. The title of William Thomson's (Lord Kelvin's) paper, in which he presented the general formulation of the second law for the first time, was: "On the Universal Tendency in Nature to the Dissipation of Mechanical Energy" (Thomson 1952). Clausius also used a cosmological language: "The entropy of the universe tends to a maximum" (Clausius 1865). However, it must be recognized that the formulation of the second law seems to us today to be more a program than a well-defined statement, because no recipe was formulated by either Thomson or Clausius to express the entropy change in terms of observable quantities. This lack of clarity in its formulation was probably one of the reasons why the application of thermodynamics became rapidly restricted to equilibrium, the end state of thermodynamic evolution. For example, the classic work of Gibbs, which was so influential in the history of thermodynamics, carefully avoids every incursion into the field of nonequilibrium processes (Gibbs 1975). Another reason may well have been that irreversible processes are nuisances in many problems: for example, they are obstacles to obtaining the maximum yield in thermal engines. Therefore the aim of engineers constructing thermal engines has been to minimize losses due to irreversible processes.

It is only recently that a complete change in perspective has arisen, and we begin to understand the *constructive* role played by irreversible processes in the physical world. Of course, the situation corresponding to equilibrium remains the simplest one. It is in this case that the entropy depends on the minimum number of variables. Let us briefly examine some classical arguments.

Consider a system that exchanges energy, but not matter, with the outside world. Such a system is called a closed system in contrast with an open system, which exchanges matter as well as energy with the outside world. Suppose that this closed system is in equilibrium. The entropy production then vanishes. On the other hand, the change of the macroscopic entropy is then defined by the heat received from the outside world. By definition,

$$d_e S = \frac{dQ}{T}, \quad d_i S = 0 \tag{4.1}$$

in which T is a positive quantity called the *absolute temperature*.

Let us combine this relation with the first law of thermodynamics, as is valid for such a simple system (for details, see Prigogine 1967):

$$dE = dQ - p\,dV \tag{4.2}$$

in which E is the energy, p the pressure, and V the volume. This formula expresses that the energy exchanged by the system with the outside world during a small time interval dt is due to the heat received by the system plus the mechanical work performed at its boundaries. Combining equation 4.1 with equation 4.2, we obtain the total differential of the entropy in the variables E and V:

$$dS = \frac{dE}{T} + p\frac{dV}{T} \tag{4.3}$$

Gibbs has generalized this formula to include variations in composition. Let us call $n_1, n_2, n_3, \ldots$, the number of moles of the various components. We may then write

$$dS = \left(\frac{dS}{\partial E}\right)dE + \frac{\partial S}{\partial V}\,dV + \sum_\gamma \left(\frac{\partial S}{\partial n_\gamma}\right)dn_\gamma$$

$$= \frac{dE}{T} + \frac{p}{T}\,dV - \sum \frac{\mu_\gamma}{T}\,dn_\gamma \tag{4.3'}$$

The quantities μ_γ are by definition the *chemical potentials* introduced by Gibbs, and equation 4.3' is called the Gibbs formula for entropy. The chemical potentials are themselves functions of the thermodynamic variables, such as temperature, pressure, concentration, and so forth. They take an especially simple form for so-called ideal systems,* in which they depend in a logarithmic way on the mole fractions $N_\gamma = n_\gamma/(\sum n_\gamma)$:

$$\mu_\gamma = \zeta_\gamma(p, T) + RT \log N_\gamma \tag{4.4}$$

in which R is the gas constant (equal to the product of Boltzmann's constant k and Avogadro's number) and $\zeta_\gamma(p, T)$ is some function of pressure and temperature.

---

* Examples of ideal systems are dilute solutions and perfect gases.

Instead of entropy, other thermodynamic potentials are often introduced, such as Helmholtz free energy, defined by

$$F = E - TS \tag{4.5}$$

It is then easy to show that the law of increase of entropy, valid for isolated systems, is replaced by the law of *decrease* of free energy for systems that are maintained at a given temperature.

The structure of equation 4.5 reflects a competition between the energy E and the entropy S. At low temperatures the second term is negligible and the minimum value of F imposes structures corresponding to minimum energy and generally to low entropy. At increasing temperatures, however, the system shifts to structures of higher and higher entropy.

Experience confirms these considerations because at low temperatures we find the solid state characterized by an ordered structure of low entropy, whereas at higher temperatures we find the gaseous state of high entropy. The formation of certain types of ordered structures in physics is a consequence of the laws of thermodynamics applied to closed systems at thermal equilibrium.

In Chapter 1, the simple interpretation of entropy in terms of complexions, given by Boltzmann, was described. Let us apply this formula to a system whose energy levels are given by E_1, E_2, E_3. By looking for the occupation numbers, which make the number of complexions (equation 1.9) a maximum for given values of the total energy and number of particles, we obtain Boltzmann's basic formula for the probability, P_i, of the occupation of a given energy level, E_i:

$$P_i = e^{-E_i/kT} \tag{4.6}$$

in which k is, as in equation 1.10, Boltzmann's constant, T the temperature, and E_i the energy of the chosen level. Suppose that we consider a simplified system with only three energy levels. Then Boltzmann's formula (equation 4.6) tells us the probability of finding a molecule in each of the three states at equilibrium. At very low temperatures, $T \to 0$, the only significant probability is that corresponding to the lowest energy level, and we come to the scheme shown in Figure 4.1 in which virtually all the molecules are in the lowest energy state, E_1, because

$$e^{-E_1/kT} \gg e^{-E_2/kT}, e^{-E_3/kT} \tag{4.7}$$

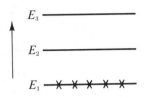

FIGURE 4.1
Low-temperature distribution: only the lowest energy
level is appreciably populated.

At high temperatures, however, the three probabilities become roughly equal:

$$e^{-E_1/kT} \simeq e^{-E_2/kT} \simeq e^{-E_3/kT} \tag{4.8}$$

and therefore the three states are approximately equally populated (see Figure 4.2).

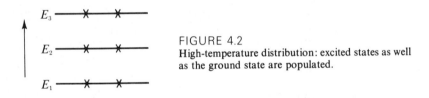

FIGURE 4.2
High-temperature distribution: excited states as well
as the ground state are populated.

Boltzmann's probability distribution (equation 4.6) gives us the basic principle that governs the structure of equilibrium states. It may appropriately be called *Boltzmann's order principle*. It is of paramount importance as it is capable of describing an enormous variety of structures including, for example, some as complex and delicately beautiful as snow crystals (Figure 4.3).

Boltzmann's order principle explains the existence of equilibrium structures. However, the question can be asked, Are they the only type of structures that we see around us? Even in classical physics we have many phenomena where nonequilibrium may lead to order. When we apply a thermal gradient to a mixture of two different gases, we observe an increment of one of the components at the hot wall, whereas the other concentrates at the cold wall. This phenomenon, already observed in the nineteenth century, is called thermal diffusion. In the steady state, the entropy is generally lower than it would be in a uniform texture. This shows that nonequilibrium may be a source of order. This observation initiated the point of view originated by the Brussels school. (See Prigogine and Glansdorff, 1971, for a historical survey.)

FIGURE 4.3
Typical snow crystals. (Courtesy of National Oceanic
and Atmospheric Administration, photographs
by W. A. Bentley.)

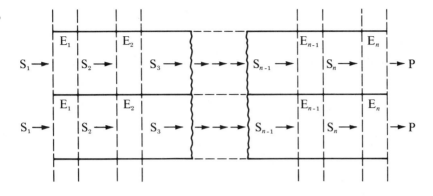

FIGURE 4.4
Mosaic model of multienzyme reaction. Substrate S_1 is changed by successive modifications to the product P by the action of "captive" enzymes.

The role of irreversible processes becomes much more marked when we turn to biological or social phenomena. Even in the simplest cells, the metabolic function includes several thousand coupled chemical reactions and, as a consequence, requires a delicate mechanism for their coordination and regulation. In other words, we need an extremely sophisticated *functional* organization. Furthermore, the metabolic reactions require specific catalysts, the enzymes, which are large molecules possessing a spatial organization, and the organism must be capable of synthesizing these substances. A catalyst is a substance that accelerates a certain chemical reaction but is not itself used up in the reaction. Each enzyme, or catalyst, performs one specific task; if we look at the manner in which the cell performs a complex sequence of operations, we find that it is organized along exactly the same lines as a modern assembly line (see Figure 4.4). (See Welch, 1977.)

The overall chemical modification is broken down into successive elementary steps, each of which is catalyzed by a specific enzyme. The initial compound is labeled S_1 in the diagram; at each membrane, an "imprisoned" enzyme performs a given operation on the substance and then sends it on to the next stage. Such an organization is quite clearly not the result of an evolution toward molecular disorder! Biological order is both architectural and functional; furthermore, at the cellular and supercellular levels, it manifests itself by a series of structures and coupled functions of growing complexity and hierarchical character. This is contrary to the concept of evolution as described in the thermodynamics of isolated systems, which leads simply to the state of maximum number of complexions and, therefore, to "disorder." Do we then have to conclude, as did Roger Caillois (1976), that "Clausius and Darwin cannot both be right," or should we introduce, with Herbert Spencer (1870), some new principle of nature, such as the "instability of the homogeneous" or "a differentiating force, creator of organization."

The unexpected new feature is that nonequilibrium may, as will be seen in this chapter, lead to a new type of structure, the *dissipative* structures, which are essential in the understanding of coherence and organization in the nonequilibrium world in which we live.

Linear Nonequilibrium Thermodynamics

To make the transformation from equilibrium to nonequilibrium, we must calculate entropy production explicitly. We can no longer be satisfied with the simple inequality, because we want to relate entropy production to well-defined physical processes. A simple evaluation of entropy production becomes possible if we *assume* that entropy outside equilibrium depends on the same variables, E, V, n_y, as it does in equilibrium. (For nonuniform systems we would have to assume that the entropy density depends on the energy density and local concentrations.) As an example, let us calculate the entropy produced by chemical reactions in closed systems. Consider a reaction such as

$$X + Y \rightarrow A + B. \tag{4.9}$$

The change due to the reaction in the number of moles of component X in time dt is equal to that of Y and opposite that of A or B:

$$dn_X = dn_Y = -dn_A = -dn_B = d\xi \tag{4.10}$$

In general, chemists introduce an integer v_y (positive or negative), called the stoichiometric coefficient of component y, into the chemical reaction; ξ is then, by definition, the degree of advancement of the chemical reaction. We may then write

$$dn_y = v_y \, d\xi \tag{4.11}$$

The rate of the reaction is

$$v = \frac{d\xi}{dt} \tag{4.12}$$

Taking into account this expression as well as the Gibbs formula (4.3'), we immediately obtain,

$$dS = \frac{dQ}{T} + \frac{A \, d\xi}{T} \tag{4.13}$$

in which A is the *affinity* of the chemical reactions (first introduced by Theophile De Donder, 1936), which is related to the chemical potentials, μ_j, by

$$A = -\sum v_\gamma \mu_\gamma \tag{4.14}$$

The first term in equation 4.13 corresponds to an entropy flow (see equation 4.1), whereas the second term corresponds to the entropy production:

$$d_i S = A \frac{d\xi}{T} \geqslant 0 \tag{4.15}$$

Using definition 4.12, we find that the entropy production per unit time takes the remarkable form

$$\frac{d_i S}{dt} = \frac{A}{T} v \geqslant 0 \tag{4.16}$$

It is a bilinear form in the rate v of the irreversible process (here the chemical reaction) and the corresponding force (here A/T). This type of calculation can be generalized: starting with the Gibbs formula (4.3'), one obtains

$$\frac{d_i S}{dt} = \sum_j X_j J_j \geqslant 0 \tag{4.17}$$

in which J_j represents the rates of the various irreversible processes taking place (chemical reactions, heat flow, diffusion, etc.) and X_j the corresponding generalized forces (affinities, gradients of temperature and of chemical potentials, etc.). This is the basic formula of the macroscopic thermodynamics of irreversible processes.

It should be emphasized that supplementary assumptions have been used to derive the explicit expression (4.17) for the entropy production. The validity of the Gibbs formula (4.3′) can only be established in some neighborhood of equilibrium. This neighborhood defines the region of "local" equilibrium.

At thermodynamic equilibrium,

$$J_i = 0, \quad X_i = 0 \qquad (4.18)$$

for *all* irreversible processes simultaneously. It is therefore quite natural to assume, at least for conditions near equilibrium, linear homogeneous relations between the flow and the forces. Such a scheme automatically includes such empirical laws as Fourier's law, which says that the flow of heat is proportional to the gradient of temperature, and Fick's law for diffusion, which states that the flow of diffusion is proportional to the gradient of concentration. In this way, we obtain the linear thermodynamics of irreversible processes characterized by the relations

$$J_i = \sum_j L_{ij} X_j \qquad (4.19)$$

Linear thermodynamics of irreversible processes is dominated by two important results. The first is expressed by the Onsager reciprocity relations (1931), which state that

$$L_{ij} = L_{ji} \qquad (4.20)$$

When the flow J_i, corresponding to the irreversible process i, is influenced by the force X_j of the irreversible process j, then the flow J_j is also influenced by the force X_i through the *same* coefficient L_{ij}.

The importance of the Onsager relations resides in their generality. They have been submitted to many experimental tests. Their validity showed, for the first time, that nonequilibrium thermodynamics leads, as does equilibrium thermodynamics, to *general results independent of any specific molecular model*. The discovery of the reciprocity relations can be considered to have been a turning point in the history of thermodynamics.

Heat conductivity in crystals affords a simple application of Onsager's theorem. According to the reciprocity relations, the heat conductivity tensor would be symmetrical whatever the symmetry of the crystal. This remarkable property had in fact already been established experimentally by Woldemaz Voigt in the nineteenth century and corresponds to a special case of the Onsager relations.

The proof of Onsager's relations is given in textbooks (see Prigogine 1967). The important point for us is that they correspond to a general property independent of any molecular model. This is the feature that makes them a thermodynamic result.

Another example to which Onsager's theorem applies is a system composed of two vessels connected by means of a capillary or a membrane. A temperature difference is maintained between the two vessels. This system has two forces, say X_k and X_m, corresponding to the difference in temperature and chemical potential between the two vessels, and two corresponding flows, J_k and J_m. It reaches a state in which the transport of matter vanishes, whereas the transport of energy between the two phases at different temperatures continues; that is, a *steady non-equilibrium state*. No confusion should arise between such states and equilibrium states characterized by a zero entropy production.

According to equation 4.17, entropy production is given by

$$\frac{d_i S}{dt} = J_k X_k + J_m X_m \qquad (4.21)$$

with the linear phenomenological laws (see equation 4.19)

$$J_k = L_{11} X_k + L_{12} X_m$$

$$J_m = L_{21} X_k + L_{22} X_m \qquad (4.22)$$

For the stationary state, the flow of matter vanishes:

$$J_m = L_{21} X_k + L_{22} X_m = 0 \qquad (4.23)$$

Coefficients L_{11}, L_{12}, L_{21}, L_{22} are all measurable quantities, and we may therefore verify that, indeed,

$$L_{12} = L_{21} \qquad (4.24)$$

This example can be used to illustrate the second important property of linear nonequilibrium systems: the theorem of minimum entropy production (Prigogine, 1945; see also Glansdorff and Prigogine, 1971). It is easy to see that equation 4.23, together with equation 4.24, is equivalent to the condition that entropy production (equation 4.21) is minimum for a given constant X_k. Equations 4.21, 4.22, and 4.24 give

$$\frac{1}{2}\frac{\partial}{\partial X_m}\left(\frac{d_i S}{dt}\right) = (L_{12} X_k + L_{22} X_m) = J_m \qquad (4.25)$$

Therefore the vanishing of the mass flow (equation 4.23) is equivalent to the extremum condition

$$\frac{\partial}{\partial X_m}\left(\frac{d_i S}{dt}\right) = 0 \qquad (4.26)$$

The theorem of minimum entropy production expresses a kind of "inertial" property of nonequilibrium systems. When given boundary conditions prevent the system from reaching thermodynamic equilibrium (i.e., zero entropy production) the system settles down in the state of "least dissipation."

It was clear when this theorem was formulated that it was strictly valid only in the neighborhood of equilibrium, and for many years great efforts were made to extend this theorem to systems farther from equilibrium. It came as a great surprise when it was shown that in systems far from equilibrium the thermodynamic behavior could be quite different—in fact, even *directly opposite* that predicted by the theorem of minimum entropy production.

It is remarkable that this unexpected behavior had already been observed in ordinary situations studied in classical hydrodynamics. The example first analyzed from this point of view is called the *Bénard instability*. (For a detailed discussion of this and other hydrodynamic instabilities, see Chandrasekhar, 1961.)

Consider a horizontal layer of fluid between two infinite parallel planes in a constant gravitational field, and let us maintain the lower boundary at temperature T_1 and the higher boundary at temperature T_2 with $T_1 > T_2$. For a sufficiently large value of the "adverse" gradient $(T_1 - T_2)/(T_1 + T_2)$, the state of rest becomes unstable and convection

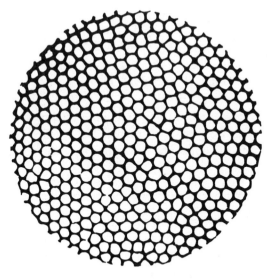

FIGURE 4.5
Spatial pattern of convection cells, viewed from above in a
liquid heated from below.

starts. Entropy production is then *increased* because the convection is a
new mechanism for heat transport (see Figure 4.5). Moreover, the motions of the currents that appear after convection has been established are
more highly organized than are the microscopic motions in the state of
rest. In fact, large numbers of molecules must move in a coherent fashion
over observable distances for a sufficiently long time for there to be a
recognizable pattern of flow.

Thus, we have a good example of the fact that nonequilibrium can be a
source of order. As will be seen later in this chapter in the section titled
Application to Chemical Reactions, this is true not only for hydrodynamic systems, but also for chemical systems if well-defined conditions
imposed upon the kinetic laws are satisfied.

It is interesting to note that Boltzmann's order principle would assign
almost zero probability to the occurrence of Bénard convection. Whenever new coherent states occur far from equilibrium, the application of
probability theory, as implied in the counting of number of complexions,
breaks down. For Bénard convection, we may imagine that there are
always small convection currents appearing as fluctuations from the average state, but below a certain critical value of the temperature gradient,
these fluctuations are damped and disappear. However, above this critical

value, certain fluctuations are amplified and give rise to a macroscopic current. A new molecular order appears that basically corresponds to a giant fluctuation stabilized by the exchange of energy with the outside world. This is the order characterized by the occurrence of what are referred to as "dissipative structures."

Before further discussion of the possibility of dissipative structures, a brief review of some aspects of thermodynamic stability theory will yield interesting information about the conditions for their occurrence.

Thermodynamic Stability Theory

The states corresponding to thermodynamic equilibrium or the steady states corresponding to a minimum of entropy production in linear non-equilibrium thermodynamics are automatically stable. The concept of Lyapounov functions was introduced in Chapter 1. Entropy production in the range of linear nonequilibrium thermodynamics is just such a function: if a system is perturbed, entropy production will increase, but the system reacts by returning to the state at which its entropy production is lowest. For a discussion of far-from-equilibrium systems, it is useful to introduce still another Lyapounov function. As we know, equilibrium states in isolated systems are stable when corresponding to the maximum production of entropy. If we perturb a system that is near an equilibrium value S_e, we obtain

$$S = S_e + \delta S + \tfrac{1}{2} \delta^2 S \qquad (4.27)$$

However, because the function S is maximum at S_e, the first-order term vanishes, and therefore the stability is determined by the sign of the second-order term $\delta^2 S$.

Elementary thermodynamics permits us to calculate this important expression explicitly. First consider the perturbation of a single, independent variable, the energy E in equation 4.3'. We then have

$$\delta S = \frac{\delta E}{T}$$

and

$$\delta^2 S = \frac{\partial^2 S}{\partial E^2} (\delta E)^2 = \frac{\partial \frac{1}{T}}{\partial E} (\delta E)^2$$

$$= - C_v \frac{(\delta T)^2}{T^2} < 0 \qquad (4.28)$$

in which we have used the fact that the specific heat is defined as

$$C_v = \left(\frac{dE}{dT} \right)_v \qquad (4.29)$$

and is a positive quantity. More generally, if we perturb all the variables in equation 4.3′, we obtain a quadratic form. The result is given below (the calculations may be found, e.g., in Glansdorff and Prigogine, 1971):

$$T \, \delta^2 S = - \left[\frac{C_v}{T} (\delta T)^2 + \frac{\rho}{X} (\delta v)_{N_j}^2 + \sum_{jj'} \mu_{jj'} \, \delta N_j \, \delta N_{j'} \right] < 0 \quad (4.30)$$

Here ρ is the density, $v = 1/\rho$ is the specific volume (the index N_j means that composition is maintained constant in the variation of N_j), X is the isothermal compressibility, N_j is the mole fraction of component j, and $\mu_{jj'}$ is the derivative

$$\mu_{jj'} = \left(\frac{\partial \mu_j}{\partial N_{j'}} \right)_{p, T} . \qquad (4.31)$$

The basic stability conditions of classical thermodynamics are

$$C_v > 0 \quad \text{(thermal stability)} \qquad (4.32)$$

$$X > 0 \quad \text{(mechanical stability)} \qquad (4.33)$$

$$\sum_{jj'} \mu_{jj'} \, \delta N_j \, \delta N_{j'} > 0 \, \text{(stability with respect to diffusion)} \qquad (4.34)$$

Each of these conditions has a simple physical meaning. For example, if condition 4.32 were violated, a small fluctuation in temperature would be amplified through the Fourier equation instead of being damped.

When these conditions are satisfied, $\delta^2 S$ is a negative definite quantity. Moreover, it can be shown that the time derivative of $\delta^2 S$ is related to the entropy production, P, through

$$\frac{1}{2}\frac{\partial}{\partial t}\delta^2 S = \sum_{\rho} J_{\rho} X_{\rho} = P \geqslant 0 \qquad (4.35)$$

in which P is defined as

$$P \equiv \frac{d_i S}{dt} \geqslant 0 \qquad (4.36)$$

As a result of inequalities 4.30 and 4.35, $\delta^2 S$ is a Lyapounov function, and its existence ensures the damping of all fluctuations. That is the reason why a macroscopic description for large systems that are near equilibrium is sufficient. Fluctuations play only a subordinate role: they appear as a negligible correction to the laws for large systems.

Can this stability be extrapolated for systems farther from equilibrium? Does $\delta^2 S$ play the role of a Lyapounov function when we consider larger deviations from equilibrium but still within the framework of macroscopic description? To answer these questions, calculate the perturbation $\delta^2 S$ for a system in a *nonequilibrium state*. Inequality 4.30 remains valid, in the range of macroscopic description. However, the time derivative of $\delta^2 S$ is no longer related to the total entropy production, as in inequality 4.35, but to the entropy production arising from perturbation. In other words, we now have, as has been shown by Glansdorff and myself (1971),

$$\frac{1}{2}\frac{\partial}{\partial t}\delta^2 S = \sum_{\rho} \delta J_{\rho} \, \delta X_{\rho} \qquad (4.37)$$

The right-hand side is what may be called the excess entropy production. It should be re-emphasized that δJ_{ρ} and δX_{ρ} are deviations from the values J_{ρ} and X_{ρ} at the stationary state, the stability of which we are testing through a perturbation. Contrary to what happens for systems at equilibrium or near equilibrium, the right-hand side of equation 4.37, corresponding to the excess entropy production, does generally not have a well-defined sign. If for all t larger than some fixed time t_0, in which t_0 may be the starting time of the perturbation, we have

$$\sum_{\rho} \delta J_{\rho} \, \delta X_{\rho} \geqslant 0 \qquad (4.38)$$

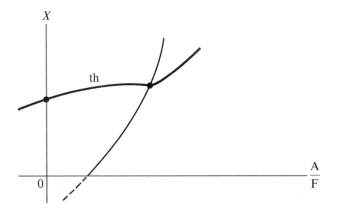

FIGURE 4.6
Various steady-state solutions corresponding to reaction 4.39:
0 corresponds to thermodynamic equilibrium; "th" is the "thermodynamic
branch."

then $\delta^2 S$ is a Lyapounov function and stability is ensured. Note that, in
the linear range, the excess entropy production would have the same sign
as the entropy production itself and we would obtain the same result as
would be obtained with the theorem of minimum entropy production.
However, the situation changes in the far-from-equilibrium range. There
the form of chemical kinetics plays an essential role.

Examples of the effect of chemical kinetics are presented in the next
section. For certain types of chemical kinetics the system may become
unstable. This shows that there is an essential difference between the laws
for systems at equilibrium and those for systems that are far from equilibrium. The laws of equilibrium are *universal*. However, far from equilibrium the behavior may become very *specific*. This is a welcome
circumstance, because it permits us to introduce a distinction in the behavior of physical systems that would be incomprehensible in an equilibrium world.

Suppose that we consider a chemical reaction of the type

$$\{A\} \rightarrow \{X\} \rightarrow \{F\} \tag{4.39}$$

in which $\{A\}$ is a set of initial products, $\{X\}$ a set of intermediate ones, and
$\{F\}$ a set of final products. Chemical reaction equations are generally
nonlinear. As a result, we shall obtain many solutions for the intermediate
concentration (see Figure 4.6). Among these solutions, *one* corresponds to

thermodynamic equilibrium and can be continued into the nonequilibrium range; it will be referred to as the *thermodynamic branch*. The
important new feature is that this thermodynamic branch may become
unstable at some critical distance from equilibrium.

Application to Chemical Reactions

Let us apply the preceding formalism to chemical reactions. Condition
4.38 for the existence of a Lyapounov function then becomes

$$\sum_{\rho} \delta v_{\rho}\, \delta A_{\rho} \geqslant 0 \tag{4.40}$$

in which δv_{ρ} represents the perturbation of the chemical reaction rates
and δA_{ρ}, the perturbation of the chemical affinities as defined in equation
4.14. Consider the chemical reaction

$$X + Y \rightarrow C + D \tag{4.41}$$

Because we are mainly interested in far-from-equilibrium situations, we
neglect the reverse reactions and write

$$v = XY \tag{4.42}$$

for the reaction rate.* According to equations 4.4 and 4.14, the affinity for
an ideal system is a logarithmic function of the concentration. Therefore,

$$A = \log \frac{XY}{CD} \tag{4.43}$$

A fluctuation in the concentration X about some steady-state value gives
rise to excess entropy production:

$$\delta v \cdot \delta A = (Y\, \delta X) \cdot \left(\frac{\delta X}{X}\right) = \frac{Y}{X}(\delta X)^2 > 0 \tag{4.44}$$

Such a fluctuation could therefore not violate stability condition 4.40.

---

* For the sake of simplicity, we will assume that all kinetic and equilibrium constants, as
well as RT, are equal to 1; also we will use X for the concentration of X, C_X, and so forth.

Let us now consider the autocatalytic reaction (instead of reaction 4.41):

$$X + Y \rightarrow 2X \tag{4.45}$$

The reaction rate is still assumed to be given by equation 4.42, but the affinity is now

$$A = \log \frac{XY}{X^2} = \log \frac{Y}{X} \tag{4.46}$$

We now have the "dangerous" contribution to excess entropy production:

$$\delta v \, \delta A = (Y \, \delta X)\left(-\frac{\delta X}{X}\right) = -\frac{Y}{X}(\delta X)^2 < 0 \tag{4.47}$$

The negative sign does not mean that the perturbed steady state will necessarily become unstable, but it may become so (the positive sign is a sufficient but not necessary condition for stability). It is, however, a general result that instability of the thermodynamic branch necessarily involves autocatalytic reactions.

One is immediately reminded of the fact that most biological reactions depend on feedback mechanisms. In Chapter 5, it will be seen, for example, that the energy-rich molecule adenosine triphosphate (ATP), necessary for the metabolism of living systems, is produced through a succession of reactions in the glycolytic cycle that involve ATP at the start. To produce ATP we need ATP. Another example is cell production: it takes a cell to produce a cell.

Thus, it becomes very tempting to relate the structure, which is so conspicuous in biological systems, to a breakdown of the stability of the thermodynamic branch. Structure and function become closely related.

To grasp this important point in a clear way let us consider some simple schemes of catalytic reactions. For example:

$$A + X \xrightarrow{k_1} 2X$$

$$X + Y \xrightarrow{k_2} 2Y$$

$$Y \xrightarrow{k_3} E \tag{4.48}$$

The values of the initial products A and final products E are maintained constant in time so that only two independent variables, X and Y, are left. To simplify, we neglect the reverse reactions. This is a scheme of auto-catalytic reactions; the increase in the concentration of X depends on the concentration of X. The same is true for Y.

This model has been widely used in ecological modelling, with X representing, for example, an herbivore that uses A, and Y representing carnivore that propagates at the expense of the herbivore. This model is associated in the literature with the names of Lotka and Volterra (see May, 1974).

Let us write the corresponding kinetic laws:

$$\frac{dX}{dt} = k_1 AX - k_2 XY \tag{4.49}$$

$$\frac{dY}{dt} = k_2 XY - k_3 Y \tag{4.50}$$

They admit a single nonvanishing steady-state solution:

$$X_0 = \frac{k_3}{k_2}; \quad Y_0 = \frac{k_1}{k_2} A \tag{4.51}$$

To study the stability of this steady state, which corresponds in this case to the thermodynamic, we shall use a normal mode analysis. We write

$$X(t) = X_0 + xe^{\omega t}; \quad Y(t) = Y_0 + ye^{\omega t} \tag{4.52}$$

with

$$\left|\frac{x}{X_0}\right| \ll 1; \quad \left|\frac{y}{Y_0}\right| \ll 1 \tag{4.53}$$

and introduce equation 4.52 into kinetic equations 4.49 and 4.50, neglecting higher-order terms in x and y. We then obtain the dispersion equation for ω (which expresses the fact that the determinant for the homogeneous linear equations vanishes). Because we have two components, X and Y, the dispersion equations are of the second order. Their explicit form is

$$\omega^2 + k_1 k_3 A = 0 \tag{4.54}$$

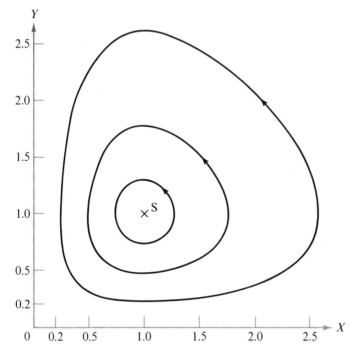

FIGURE 4.7
Periodic solutions of the Lotka-Volterra model obtained for different values
of the initial conditions.

Obviously, stability is related to the sign of the real parts of the roots of
the dispersion equations. If for each solution, ω_n, of the dispersion
equation

$$\text{re } \omega < 0 \qquad (4.55)$$

the initial state would be stable. In the Lotka-Volterra case, the real part
vanishes, and we obtain

$$\text{re } \omega_n = 0, \quad \text{im } \omega_n = \pm(k_1 k_3 A)^{1/2} \qquad (4.56)$$

This means that we have so-called marginal stability. The system rotates
around the steady state (equation 4.51). The frequency of rotation (equation 4.56) corresponds to the limit of small perturbations. The frequency
of oscillations is amplitude dependent and there are an infinite number of
periodic orbits around the steady state (see Figure 4.7).

Let us consider another example, which has been used extensively recently because it has remarkable properties that allow one to model a wide range of macroscopic behaviors. It is called *Brusselator*, and it corresponds to the scheme of reaction (for details, see Nicolis and Prigogine, 1977)

$$A \rightarrow X \qquad \text{(a)}$$

$$2X + Y \rightarrow 3X \qquad \text{(b)}$$

$$B + X \rightarrow Y + D \qquad \text{(c)}$$

$$X \rightarrow E \qquad \text{(d)} \qquad (4.57)$$

The initial and final products (A, B, C, D, and E) remain constant, whereas the two intermediate components (X and Y) may have concentrations that change in time. Putting the kinetic constants equal to one, we obtain the system of equations

$$\frac{dX}{dt} = A + X^2 Y - BX - X \qquad (4.58)$$

$$\frac{dY}{dt} = BX - X^2 Y \qquad (4.59)$$

which admits the steady state

$$X_0 = A, \quad Y_0 = \frac{B}{A} \qquad (4.60)$$

Applying the normal mode analysis, as for the Lotka-Volterra example, we obtain the equation

$$\omega^2 + (A^2 - B + 1)\omega + A^2 = 0 \qquad (4.61)$$

which should be compared with equation 4.54.

One finds immediately that the real part of one of the roots becomes positive whenever

$$B > 1 + A^2 \qquad (4.62)$$

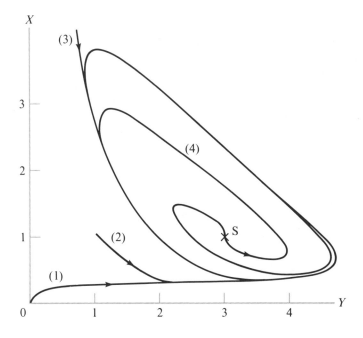

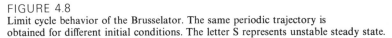

FIGURE 4.8
Limit cycle behavior of the Brusselator. The same periodic trajectory is
obtained for different initial conditions. The letter S represents unstable steady state.

Therefore, this scheme, contrary to what happens with the Lotka-
Volterra equation, presents a real instability. Numerical calculations as
well as analytical work performed for values of B, larger than the critical
value, lead to the behavior indicated in Figure 4.8. We have now a *limit
cycle*; that is, any initial point in the space XY approaches the same
periodic trajectory in time. It is important to note the very unexpected
character of this result. The oscillation frequency now becomes a
well-defined function of the physico-chemical state of the system, whereas,
in the Lotka-Volterra case, the frequency is essentially arbitrary (because
it is amplitude dependent).

Today many examples of oscillating systems are known, especially in
biological systems, and the important feature is that their oscillation
frequency is well defined once the state of the system is given. This shows
that these systems are beyond the stability of the thermodynamic branch.
Chemical oscillations of this type are *supercritical* phenomena. The
molecular mechanism leads to quite fascinating and difficult questions to
which we shall return in Chapter 6.

Limit cycle is not the only possible type of supercritical behavior. Suppose that we consider the exchange of matter between two boxes (referred to as Box 1 and Box 2). Instead of obtaining equations 4.58 and 4.59, we obtain

$$\frac{dX_1}{dt} = A + X_1^2 Y_1 - BX_1 - X_1 + D_X(X_2 - X_1),$$

$$\frac{dY_1}{dt} = BX_1 - X_1^2 Y_1 + D_Y(Y_2 - Y_1),$$

$$\frac{dX_2}{dt} = A + X_2^2 Y_2 - BX_2 - X_2 + D_X(X_1 - X_2),$$

$$\frac{dY_2}{dt} = BX_2 - X_2^2 Y_2 + D_Y(Y_1 - Y_2) \tag{4.63}$$

The first two equations refer to Box 1, the last two to Box 2. Numerical calculations show that, under suitable conditions beyond the critical value, the thermodynamic state corresponding to identical concentrations of X and Y given by (see equations 4.60)

$$X_i = A, \quad Y_i = \frac{B}{A} \quad (i = 1, 2) \tag{4.64}$$

becomes unstable. An example of this behavior, as recorded by computer, is given in Figure 4.9.

We have here a *symmetry-breaking* dissipative structure. If a steady state $X_1 > X_2$ is possible, the symmetrical one corresponding to $X_2 > X_1$ is also possible. Nothing in the macroscopic equations indicates which state will result.

It is important to note that small fluctuations can no longer reverse the configurations. Once established, the symmetry-broken systems are stable. The mathematical theory of these remarkable phenomena is discussed in Chapter 5. In concluding this chapter, emphasis is placed on *three aspects* that are always linked in dissipative structures: the *function*, as expressed by the chemical equations; the *space-time structure*, which results from the instabilities; and the *fluctuations*, which trigger the in-

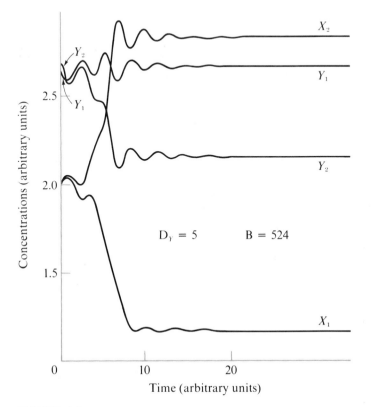

FIGURE 4.9
A perturbation of Y in Box 2 (Y_2) about the homogeneous state increases
the rate of production of X in that box (X_2), owing to the autocatalytic
step. This effect grows until a new state is reached corresponding to
spatial symmetry breaking.

stabilities. The interplay between these three aspects

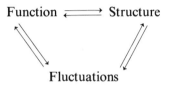

leads to the most unexpected phenomena, including *order through fluctuations*, which will be analyzed in the next two chapters.

Chapter 5

SELF-ORGANIZATION

Stability, Bifurcation,
and Catastrophes

As indicated in the preceding chapter, thermodynamic description takes various forms according to the distance from equilibrium. Of particular importance for us is the fact that, far from equilibrium, chemical systems that include catalytic mechanisms may lead to dissipative structures. As will be shown, these structures are very sensitive to global features such as the size and form of the system, the boundary conditions imposed on its surface, and so forth. All these features influence in a decisive way the type of instabilities that lead to dissipative structures. In some cases, the influence of external conditions may be even stronger; for example, macroscopic fluctuations may lead to new types of instabilities.

Far from equilibrium, therefore, an unexpected relation exists between chemical kinetics and the *space-time structure* of reacting systems. It is true that the interactions, which determine the values of the relevant kinetic constants and transport coefficients, result from short-range inter-actions (such as valency forces, hydrogen bonds, and Van der Waals forces). However, the solutions of the corresponding equations depend, *in addition*, on global characteristics. This dependence (which on the ther-modynamic branch, near equilibrium, is rather trivial) becomes decisive in chemical systems working under far-from-equilibrium conditions. For example, the occurrence of dissipative structures generally requires that the system's size exceed some critical value—a complex function of the parameters describing the reaction-diffusion processes. Therefore we may say that chemical instabilities involve *long-range order* through which the system acts *as a whole*.

This global behavior greatly modifies the very meaning of space and time. Much of geometry and physics is based on a simple concept of space and time, generally associated with Euclid and Galileo. In this view, time is homogeneous. Time translations may have no effect on physical events. Similarly, space is homogeneous and isotropic; again translations and rotations cannot alter the description of the physical world. It is quite remarkable that this simple conception of space and time may be broken by the occurrence of dissipative structures. Once a dissipative structure is formed, the homogeneity of time, as well as space, may be destroyed. We come much nearer to Aristotle's "biological" view of space-time, which was described briefly in the preface.

The mathematical formulation of these problems requires the study of partial differential equations if diffusion is taken into account. The evolu-tion of component X_i is then given by equations of the form

$$\frac{\partial X_i}{\partial t} = v(X_1, X_2, \ldots) + D_i \frac{\partial^2 X_i}{\partial r^2} \tag{5.1}$$

in which the first contribution comes from the chemical reactions and generally has a simple polynomial form (as in Chapter 4 in the section on application to chemical reactions), whereas the second term expresses diffusion along the coordinate r. For simplicity of notation we use a single coordinate r, although, in general, diffusion occurs in three-dimensional geometrical space. These equations must be supplemented by boundary conditions (generally either the concentrations or the flows are given on the boundaries).

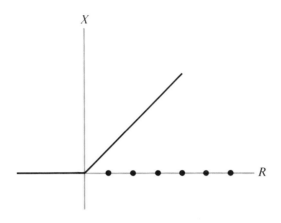

FIGURE 5.1
Bifurcation diagram for equation 5.2. Solid line and dots denote stable
and unstable branches respectively.

The variety of phenomena that may be described by this sort of
reaction-diffusion equation is quite amazing, which is why it is interesting
to consider the "basic solution" to be the one corresponding to the
thermodynamic branch. Other solutions may then be obtained by successive instabilities, which take place when the distance from equilibrium is
increased. Such types of instabilities may be studied by means of *bifurcation theory* (see Nicolis and Prigogine, 1977). In principle, a bifurcation is
simply the appearance of a new solution of the equations for some critical
value. Suppose, for example, that we have a chemical reaction corresponding to the rate equation (see McNeil and Walls, 1974)

$$\frac{dX}{dt} = \alpha X(X - R) \tag{5.2}$$

Clearly, for $R < 0$, the only time-independent solution is $X = 0$. At the
point $R = 0$, we have a bifurcation of a new solution, $X = R$ (see Figure
5.1), and it may be verified by the linear stability method explained in
Chapter 4, in the section on application to chemical reactions, that the
solution $X = 0$ then becomes unstable, whereas the solution $X = R$ becomes stable. Generally, we have successive bifurcations where we increase the value of some characteristic parameter p (like B in the
Brusselator scheme). Figure 5.2 shows a single solution for the value p_1,
but multiple solutions for the value p_2.

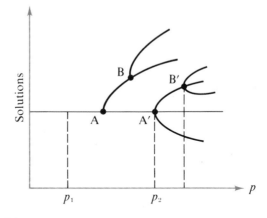

FIGURE 5.2
Successive bifurcations: A and A′ represent primary bifurcation points from the thermodynamic branch; B and B′ represent secondary bifurcation points.

It is interesting that, in a sense, the bifurcation introduces *history* into physics and chemistry, an element that formerly seemed to be reserved for sciences dealing with biological, social, and cultural phenomena. Suppose that observation shows us that the system whose bifurcation is shown in Figure 5.2 is in state C because of an increase of the value of p. Interpretation of state C implies a knowledge of the history of the system, which had to go through bifurcation points A and B.

Every description of a system that has bifurcations will imply both deterministic and probablistic elements. As will be seen in greater detail in Chapter 6, the system obeys deterministic laws, such as the laws of chemical kinetics, between two bifurcation points, but in the neighborhood of the bifurcation points fluctuations play an essential role and determine the "branch" that the system will follow. The mathematical theory of bifurcation is generally very complex. It often implies rather tedious expansions, but there are some cases in which exact solutions are available. A very simple situation of this type is provided by René Thom's (1975) theory of catastrophes, which can be applied when diffusion is neglected in equation 5.1 and when such equations derive from a potential. It means that they then take the form

$$\frac{dX_i}{dt} = -\frac{\partial V}{\partial X_i} \tag{5.3}$$

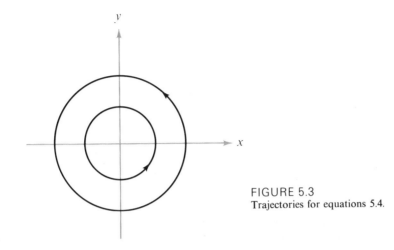

FIGURE 5.3
Trajectories for equations 5.4.

in which V is a kind of "potential function." This is a rather exceptional case. However, when satisfied, a general classification of the solutions of equation 5.3 may be undertaken by looking for the points at which there are changes in the stability properties of the steady states. These are the points that Thom called "ensemble de catastrophes."

Another type of system admitting an exact theory of bifurcation is described later in this chapter in the section titled A Solvable Model for Bifurcation.

Finally, a general concept, which plays an important role in the theory of self-organization, is that of *structural stability*. It can be illustrated by a simplified form of the Lotka-Volterra equations corresponding to the prey-predator competition:

$$\frac{dx}{dt} = by, \quad \frac{dy}{dt} = -bx \tag{5.4}$$

In the (x, y) phase space, an infinite set of closed trajectories surrounds the origin (see Figure 5.3). Compare the solutions of equations 5.4 with those arising from the following equations:

$$\frac{dx}{dt} = by + ax, \quad \frac{dy}{dt} = -bx + ay \tag{5.5}$$

In the latter case, even for the smallest value of the parameter a $(a < 0)$, the point $x = 0$, $y = 0$ is asymptotically stable, being the end point

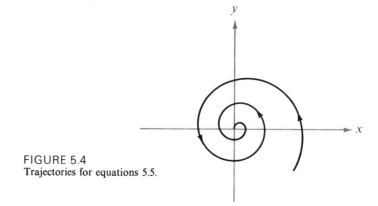

FIGURE 5.4
Trajectories for equations 5.5.

toward which all trajectories in phase space converge, as indicated in
Figure 5.4. By definition, equations 5.4 are termed "structurally
unstable" with respect to "fluctuations" that slightly alter the mechanism of interaction between x and y and introduce terms, however small, of
the type shown in equations 5.5.

This example may seem somewhat artificial, but consider a chemical
scheme describing some polymerization process in which polymers are
constructed from molecules A and B, which are pumped into the system.
Suppose that the polymer has the following molecular configuration:

$$ABABAB \ldots$$

Suppose that the reactions producing this polymer are autocatalytic. If an
error occurs and a modified polymer appears such as

$$ABAABBABA \ldots$$

then it may multiply in the system as a result of the modified autocatalytic
mechanism. Manfred Eigen presented interesting models that include
such features and showed in idealized cases that the system would evolve
toward an optimum stability with respect to the occurrence of errors in
the replication of polymers (see Eigen and Winkler, 1975). His model has
as its basis the idea of "cross catalysis." Nucleotides produce proteins,
which in turn produce nucleotides:

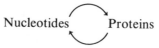

Nucleotides   Proteins

This results in a cyclic network of reactions called a *hypercycle*. When
such networks compete with one another, they display the ability to

evolve through mutation and replication into greater complexity. In recent work, Manfred Eigen and Peter Schuster (1978) presented a model for a "realistic hypercycle" related to the molecular organization of a primitive replication and translation apparatus.

The concept of structural stability seems to express in the most compact way the *idea of innovation*, the appearance of a new mechanism and a new species, which were initially absent in the system. Simple examples are given later in this chapter in the section on ecology.

Bifurcations: The Brusselator

The Brusselator model was introduced in Chapter 4. It is of interest because it presents a variety of solutions (limit cycles, nonuniform steady states, chemical waves) that are precisely of the type observed in ordinary systems that are at sufficient distance from equilibrium. When diffusion is included, the reaction-diffusion equation for the Brusselator takes the form (see equations 4.58 and 4.59 and, for details, see Nicolis and Prigogine, 1977)

$$\frac{\partial X}{\partial t} = A + X^2 Y - BX - X + D_X \frac{\partial^2 X}{\partial r^2},$$

$$\frac{\partial Y}{\partial t} = BX - X^2 Y + D_Y \frac{\partial^2 Y}{\partial r^2} \tag{5.6}$$

Suppose that we impose the value of the concentrations at the boundary. We look then for solutions of the form (see equation 4.60)

$$X = A + X_0(t) \sin \frac{n\pi r}{L},$$

$$Y = \frac{B}{A} + Y_0(t) \sin \frac{n\pi r}{L} \tag{5.7}$$

in which n is an integer and X_0 and Y_0 are still time dependent. These solutions satisfy the boundary conditions $X = A$ and $Y = B/A$ for $r = 0$

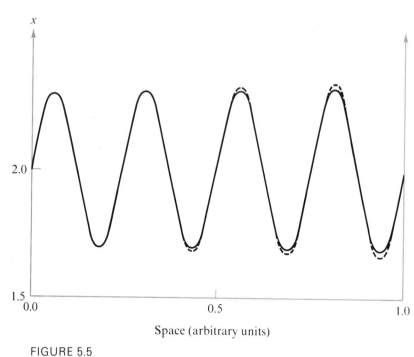

FIGURE 5.5
Steady-state dissipative structure: solid line indicates results obtained by
calculation; dashed line indicates those obtained by computer simulation
with parameters $D_x = 1.6 \times 10^{-3}$; $D_y = 8 \times 10^{-3}$; $A = 2$; $B = 4.17$.

and $r = L$. We may then apply the linear stability analysis and obtain a
dispersion equation that relates ω to the space dependence given by the
integer n in equations 5.7.

The results are as follows. The instability may arise in different ways.
The two dispersion equations may have two roots that are complex
conjugate, and at some point the real part of these roots vanishes. This is
the situation that leads to a limit cycle, which was studied in Chapter 4.
In the literature, this is often called the *Hopf bifurcation* (Hopf 1942). A
second possibility is that we have two real roots, one of which becomes
positive at some critical point. That is the situation leading to spatially
nonuniform steady states. We may call it the *Turing bifurcation*, because
Alan Turing (1952) was the first to note the possibility of such a bifurcation
in chemical kinetics in his classic paper on morphogenesis.

The variety of phenomena is even larger because the limit cycle may
also be space dependent and lead to chemical waves. Figure 5.5 shows a
chemical nonuniform steady state corresponding to a Turing bifurcation,
whereas Figure 5.6 shows, the simulation of a chemical wave. The realiza-

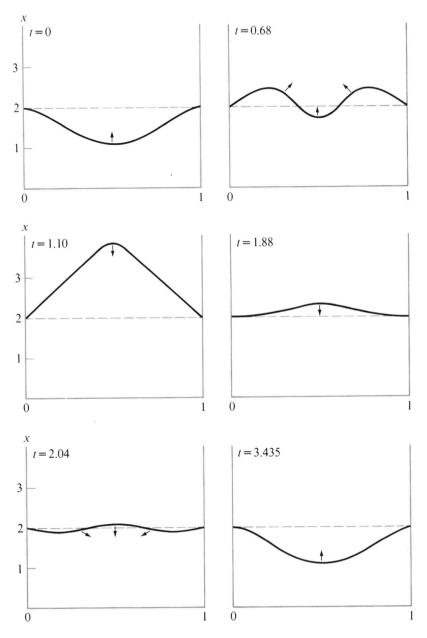

FIGURE 5.6
A chemical wave simulated on computer with parameters
$D_x = 8 \times 10^{-3}$; $D_y = 4 \times 10^{-3}$; $A = 2$; $B = 5.45$.

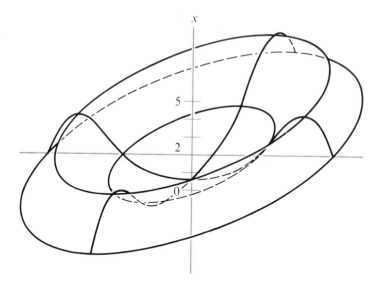

FIGURE 5.7
Cylindrically symmetric steady-state dissipative structure in two dimensions
obtained by computer simulation with parameters $D_x = 1.6 \times 10^{-3}$;
$D_y = 5 \times 10^{-3}$; $A = 2$; $B = 4.6$; circle radius $R = 0.2$.

tion of one or the other of these coherent phenomena depends on the
value of the diffusion coefficients D or, better, on the ratio D/L^2. When this
parameter becomes zero, we obtain a limit cycle, the "chemical clock,"
whereas inhomogeneous steady states can appear only when D/L^2 is
sufficiently large.

Localized structures can also result from this scheme of reactions when
the fact that the initial substances A and B (see equation 4.57) must
diffuse through the system is taken into account.

The wealth of dissipative structures increases greatly when two- or
three-dimensional systems are considered. For example, we may then
have the appearance of polarity in a hitherto uniform system. Figures 5.7
and 5.8 show the first bifurcation in a two-dimensional circular system
differing in the values of the diffusion constants. In Figure 5.7, the concentration remains radially isotropic, whereas, in Figure 5.8, the appearance
of a privileged access can be observed. This is of interest for the application to morphology in which one of the first stages corresponds to the
appearance of a gradient in a system that was initially in a spherically
symmetric state.

Successive bifurcations may also be of interest, for example, as il-

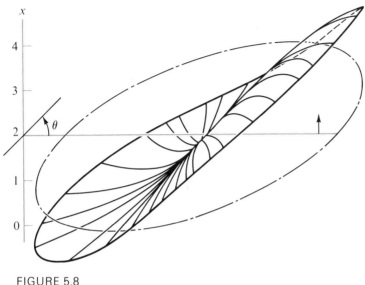

FIGURE 5.8
Polar steady-state dissipative structure in two dimensions obtained by computer simulation with parameters $D_x = 3.25 \times 10^{-3}$; $D_y = 1.62 \times 10^{-2}$; $A = 2$; $B = 4.6$; $R = 0.1$.

lustrated in Figure 5.9. Before B_0 we have the thermodynamical branch, whereas at B_0 a limit cycle behavior begins. The thermodynamical branch remains unstable but bifurcates into two new solutions at point B_1; these solutions are also unstable but become stable at points B_{1a}^*, B_{1b}^*. These two new solutions correspond to chemical waves.

One type of wave admits a plane of a symmetry (Figure 5.10), whereas the other corresponds to rotating waves (see Figure 5.11). It is quite

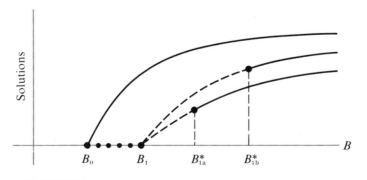

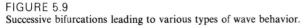

FIGURE 5.9
Successive bifurcations leading to various types of wave behavior.

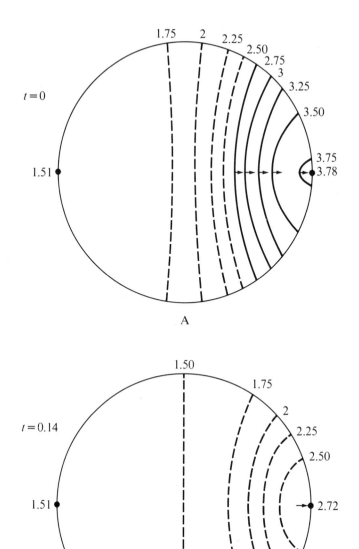

FIGURE 5.10
Equal concentration curves for X in trimolecular model in circle of radius $R = 0.5861$ subject to zero-flux boundary conditions. Solid and dashed lines refer, respectively, to concentrations larger or smaller than values on (unstable) steady state: $X_0 = 2$, $A = 2$, $D_1 = 8 \times 10^{-3}$, $D_2 = 4 \times 10^{-3}$, $B = 5.4$. The concentration patterns shown in Parts A and B are at different stages of periodic solution.

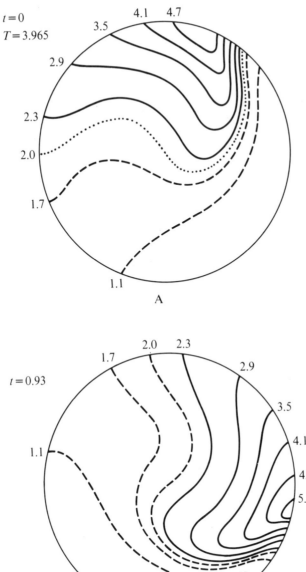

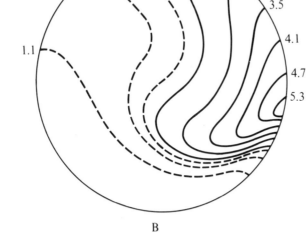

FIGURE 5.11
Rotating solution for trimolecular model arising under the same conditions as those for Figure 5.10 but for an even larger supercritical value of bifurcation parameter, $B = 5.8$.

remarkable that this type of situation has indeed been observed experimentally in chemical reactions (see the section titled Coherent Structures
in Chemistry and Biology later in this chapter).

A Solvable Model for Bifurcation

The occurrence of nonuniform, stable solutions after a bifurcation is such
an unexpected phenomenon that it is worthwhile examining their formation in an exactly solvable model (see Lefever, Herschkowitz-Kaufman,
and Turner, 1977). Consider a chemical system described by the scheme
of reaction

$$\frac{\partial X}{\partial t} = v(X, Y) + D_X \frac{\partial^2 X}{\partial z^2},$$

$$\frac{\partial Y}{\partial t} = -v(X, Y) + D_Y \frac{\partial^2 Y}{\partial z^2} \tag{5.8}$$

We may, for example, consider

$$v(X, Y) = X^2 Y - BX \tag{5.9}$$

which is the simplified form of the Brusselator in which the reaction
$A \to X$ in equation 4.57 has been suppressed. Such a description appears
in the theory of dissipative structures for reactions involving enzymes
fixed on membranes where the presence of component X is ensured by
diffusion and not through the "source" A.

Let us also use fixed boundary conditions:

$$X(0) = X(L) = \xi$$

$$Y(0) = Y(L) = \frac{B}{\xi} \tag{5.10}$$

The specific simplifying feature of this scheme is the existence of a "conserved quantity," as can be seen by adding the two equations in reaction

5.8. After eliminating one of the variables and integrating, we obtain the equation valid at the steady state

$$\left(\frac{dw}{dr}\right)^2 = K - \Phi(w) \tag{5.11}$$

in which K is the integration constant and

$$w = x - \xi \tag{5.12}$$

$\Phi(w)$ is a polynomial in w whose exact form is of no interest here. Note only that $\Phi(w) = 0$ for $w = 0$. It is very interesting to compare this formula to the Hamiltonian as expressed in equation 2.1 or 2.2, which is written here as

$$\frac{m}{2}\left(\frac{dq}{dt}\right)^2 = H - V(q) \tag{5.13}$$

It can be seen that, to change the Hamiltonian (equation 5.13) into equation 5.11, we must replace the coordinate q by the concentration and time by the coordinate r. Note also that $w = 0$ is at the boundary of the system.

Consider the systems represented in Figures 5.12 and 5.13. In the situation shown in Figure 5.12 in which $\Phi(w)$ has a maximum for $w = 0$, only the thermodynamical branch can be stable. Suppose that we start at $w = 0$ and go to the right. $\Phi(w)$ becomes negative, which means according to equation 5.11 that the gradient $(dw/dr)^2$ will increase steadily as distance from the boundary increases. We shall therefore be able to satisfy the second boundary condition.

The situation changes completely when we consider the case in which $\Phi(w)$ has a minimum for $w = 0$. Then by going to the right, we come to the point of intersection with the horizontal K. At this point, w_m, the gradient (dw/dr) will vanish, and we can then reach the second boundary by going back to the origin $w = 0$. In this way we obtain a bifurcating solution with a single extremum.

Certainly other, more complicated solutions can be built in the same way. This provides us, I believe, with the simplest, effective construction of a bifurcating solution in a reaction-diffusion system. It is interesting

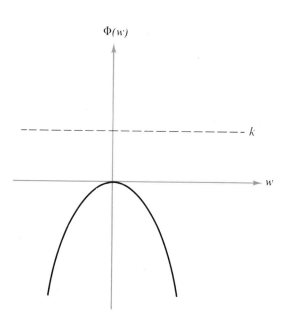

FIGURE 5.12
Situation corresponding to no bifurcation.

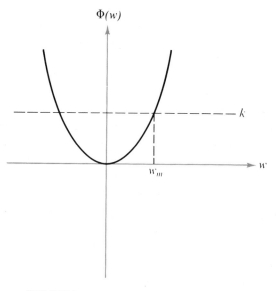

FIGURE 5.13
Situation corresponding to bifurcation.

that the time periodicity of the classical pendulum problem leads to the space periodicity of bifurcating solutions.

A further fascinating analogy between time-periodic and time-independent but spatially nonuniform solutions of nonlinear systems can be drawn by choosing as the bifurcation parameter a characteristic length, L, of the reaction space. As it turns out, if L is small enough, only the spatially homogeneous state will exist and be stable for natural boundary conditions. Above a critical value, L_{c_1}, however, a stable monotonic gradient of the kind shown in Figure 5.8 can emerge and subsist until a second critical value L'_{c_1} is reached, whereupon this pattern disappears (Babloyantz and Hiernaux 1975). The existence of this *finite* length L_{c_1} for spatial self-organization is to be compared with the emergence of a *finite* frequency accompanying the bifurcation of a time-periodic solution like a limit cycle (see the soluble model just discussed).

If L increases further, at a certain L_{c_2} ($L_{c_2} > L_{c_1}$ but possibly $< L'_{c_1}$), a second pattern will be available that will give a nonmonotonic concentration profile. Further growth will reveal more complex concentration patterns. Their relative stability will depend on the occurrence of secondary and higher bifurcations.

The fact that growth and morphology are linked in this picture is reminiscent of some aspects of morphogenesis in early embryonic development. For example, the "imaginal discs" in the early stages of larval development of the fruitfly *Drosophila* both grow and subdivide into compartments separated by rather sharp boundaries. This problem was recently analyzed by Stuart Kauffman and co-workers (1978) in the context of repeated bifurcations at successively higher lengths, as discussed above.

The existence of a second bifurcation parameter L, in addition to the kinetic bifurcation parameter p (see Figure 5.2) or B (see Figure 5.9) that is present in most systems, enables some systematic, if preliminary, classification of spatially nonuniform dissipative structures. In diagrams such as Figure 5.14, in which the structure of bifurcation is given in terms of a single parameter, only primary bifurcating branches are represented. In the vicinity of the bifurcation points their behavior is known. In particular, the first branch is stable (if supercritical, i.e., if it arises for $p > p_{c_1}$) whereas the others are unstable. Higher bifurcating branches are not shown because they typically arise at a finite distance from the bifurcation points.

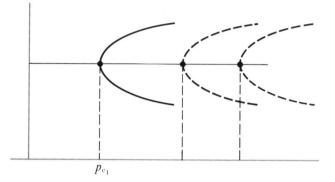

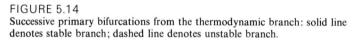

FIGURE 5.14
Successive primary bifurcations from the thermodynamic branch: solid line
denotes stable branch; dashed line denotes unstable branch.

The situation changes if bifurcation is followed in the space of *both p*
and *L*. For certain combinations of *p* and *L*, *degenerate bifurcation* at a
double eigenvalue of the linearized operator may occur, in which case
bifurcating branches coalesce. Conversely, if *p* and *L* change slightly from
this state of degeneracy, the bifurcation branches split and may give rise
to secondary and higher bifurcations (Erneux and Hiernaux, in press;
Golubitsky and Schaeffer 1979). The point is that all these possibilities can
be classified *completely*, as long as one remains near the degenerate bifurcation. The situation begins to look like catastrophe theory, even
though one is not dealing in general with systems deriving from a potential.

Coherent Structures in Chemistry and Biology

In 1958 Belousov reported an oscillating chemical reaction corresponding to the oxidation of citric acid by potassium bromate catalyzed by the
ceric-cerous ion couple. Zhabotinskii pursued this study (see Noyes and
Field, 1974, and Nicolis and Prigogine, 1977). Usually the Belousov-
Zhabotinskii reaction involves a reaction mixture at about 25°C, consisting of potassium bromate, malonic or bromomalonic acid, and ceric
sulfate or an equivalent compound dissolved in citric acid. This reaction

has been studied by many investigators both experimentally and theoretically. In experimental studies, it plays the same role as does the Brusselator in theoretical studies. According to the circumstances, a wide range of phenomena has been observed; for example, oscillations for a period of the order of a minute in homogeneous mixtures and wavelike activity. Elucidation of the mechanism of this reaction is largely attributed to Richard Noyes and co-workers (Noyes and Field 1974). Let

$$X = [HBrC_2]$$

$$Y = [Br^-]$$

$$Z = 2[Ce^{4+}] \tag{5.14}$$

be the concentrations of the three key substances. Moreover, we set

$$A = B = [BrO_3^-]$$

$$P, Q = \text{waste-product concentration} \tag{5.15}$$

The Noyes mechanism can then be expressed by the following steps:

$$A + Y \xrightarrow{k_1} X$$

$$X + Y \xrightarrow{k_2} P$$

$$B + X \xrightarrow{k_{3,4}} 2X + Z$$

$$2X \xrightarrow{k_5} Q$$

$$Z \xrightarrow{k_6} fY \tag{5.16}$$

This is often called the *Oregonator*. The important point is the existence of a cross-catalytic mechanism by which Y produces X, X produces Z, and Z in turn produces Y, just as in the Brusselator.

Many other oscillating reactions of the same type have been investigated. An early example is the catalytic decomposition of hydrogen peroxide by the iodic acid and iodine oxidation couple (see Bray, 1921, and Sharma and Noyes, 1976). More recently, Thomas Briggs and

Warren Rauscher (1973) reported oscillations in a reaction that included hydrogen peroxide, malonic acid, potassium iodate, manganous sulfate, and perchloric acid, which may be viewed as a "mixture" of Belousov-Zhabotinskii and Bray reagents. This reaction was studied systematically by Adolf Pacault and co-workers under open-systems conditions (Pacault, de Kepper, and Hanusse 1975). Finally, Endre Körös (1978) reports a whole family of simple aromatic compounds (phenol, aniline, and their derivatives), which in reacting with acid bromate are capable of generating oscillations without the catalytic action of metal ions like cerium or manganese ions. These metal ions are known to play an important role in the Belousov-Zhabotinskii reaction. Although oscillating reactions are rather exceptional in the field of inorganic chemistry, they are observed at all levels of biological organization—from the molecular to the supercellular.

Among the most significant biological oscillations are those related to enzyme activity in metabolism, which have a period of the order of a minute, and those related to epigenesis, which have a period of the order of an hour. The best understood example of metabolic oscillation is that which occurs in the glycolytic cycle, which is a phenomenon of the greatest importance for the energetics of living cells (Goldbeter and Caplan 1976). It consists in the degradation of one molecule of glucose and the overall production of two molecules of ATP by means of a linear sequence of enzyme-catalyzed reactions. It is the cooperative effects involved in the enzyme activity that lead to the catalytic effects responsible for the oscillations. It is quite remarkable that oscillations in the concentrations of all metabolites of the chain are observed for certain rates of glycolytic substrate injection. Even more remarkable is the fact that all glycolytic intermediates oscillate with the same period but with different phases. The enzymes in the reaction play somewhat the same role as a nicol prism in optical experiments. They lead to a phase shift in the chemical oscillation. The oscillatory aspect of chemical reaction is especially spectacular in the glycolytic cycle because one can follow experimentally the influence of various factors on the period and the phase of the oscillation.

Oscillatory reactions of the epigenetic type are also well known. They occur as a consequence of regulatory processes at the cellular level. Proteins are generally stable molecules, whereas catalysis is a very fast process. Thus, it is not unusual for the protein level in a cell to be too high, in which case other bodily substances act to suppress the synthesis of macromolecules. Such feedback gives rise to oscillations and has been studied in detail in, for example, the regulation of the lactose operon in the bacterium *Escherichia coli*. Other examples of oscillation-producing feedback mechanisms can be found in the aggregation process in slime molds, in reactions involving membrane-bound enzymes, and so forth. The interested reader should consult the relevant literature (for references, see Nicolis and Prigogine, 1977).

It seems that most biological mechanisms of action show that life involves far-from-equilibrium conditions beyond the stability of the threshold of the thermodynamic branch. It is therefore very tempting to suggest that the origin of life may be related to successive instabilities somewhat analogous to the successive bifurcations that have led to a state of matter of increasing coherence.

Ecology

Let us turn to some aspects of stability theory applicable to structural stability (see Prigogine, Herman, and Allen, 1977.) To take a simple example, the growth of a population X in a given medium is often expressed by

$$\frac{dX}{dt} = KX(N - X) - dX \qquad (5.17)$$

in which K is related to the rate of birth, d is related to mortality, and N is a measure of the milieu's capacity to support the population. The solution to equation 5.17 can be expressed with the help of the logistic curve presented in Figure 5.15. This evolution is entirely deterministic. The population ceases to grow when the milieu is saturated. However, it may happen, following events over which the model has no control, that a new species (characterized by other ecological parameters K, N, and d) appears, initially in a small quantity, in the same milieu. This *ecological fluctuation* raises the question of structural stability: the new species may either disappear or supplant the original one. It is easy to show, using linear stability analysis, that the new species will supplant the original

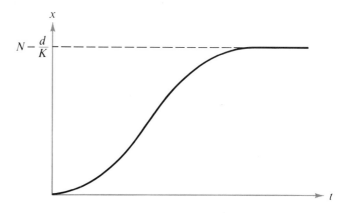

FIGURE 5.15
Logistic curve. See equation 5.17.

one only if

$$N_2 - \frac{d_2}{K_2} > N_1 - \frac{d_1}{K_1} \tag{5.18}$$

The occupation of the ecological niche by the species assumes the form indicated in Figure 5.16.

This model describes in a quantitatively exact manner the significance of "the survival of the fittest" in the framework of a problem posed in terms of the exploitation of a given ecological niche.

A variety of such models may be introduced by taking into account various possible strategies used by a population for its survival. For example, we may distinguish between species using a wide variety of foods (so-called generalists) and those using a narrow spectrum (so-called specialists). We may also take into account the fact that some populations immobilize a part of their society for "unproductive" functions, such as "soldiers." This is closely related to the social polymorphism of insects.

One could also make use of the concepts of structural stability and order through fluctuation in more complex problems, and, at the cost of some drastic simplifications, even in the study of human evolution. As an example, let us consider the problem of urban evolution from this point of view (see Allen, 1977). In terms of the logistic equation (5.17), an urban region is characterized by the increase of its capacity N because of the

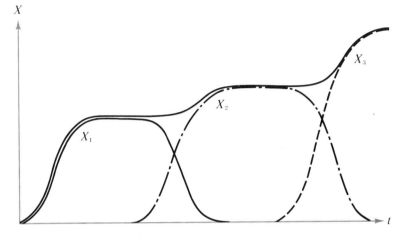

FIGURE 5.16
Occupation of an ecological niche by the successive species.

addition of economic functions. Let S_i^k be the economic function k at point i (say, the "city" is i). We must then replace equation 5.17 with an equation of the type

$$\frac{dX_i}{dt} = KX_i\left(N + \sum_k R^k S_i^k - X_i\right) - dX_i \qquad (5.19)$$

in which R^k is a coefficient of proportionality. However, S_i^k itself increases with the population X_i in a complicated manner: it plays an autocatalytic role, but the efficiency of this autocatalysis depends on the *need* at point i for the product k as related to the increase of the population and the *competition* from rival production units, located at other points.

In this model, the appearance of an economic function is comparable to a fluctuation. The appearance of this economic function will destroy the initial uniformity of the population distribution by creating employment opportunities that concentrate the population at a point. The new employment opportunities will drain the demand from neighboring points; intervening in an already urbanized area, they may be destroyed by the competition from similar but better developed or better situated economic functions; they may also develop in coexistence, or at the cost of the destruction of one or the other of these functions.

Figure 5.17 illustrates a possible "history" of the urbanization of an initially uniform region, in which four economic functions seek to develop at each point in a network of fifty localities; the various attempts follow each other in a stochastic temporal sequence. The final result depends in a complex manner on the interplay of deterministic economic laws and the probabilistic succession of fluctuations. Although the details of any particular simulation depend on the exact "history" of the region, certain average properties of the structure engendered are roughly conserved. For example, the number and average separation of large centers is approximately the same for systems having the same values of the parameters even though they undergo different histories. Such a model permits an estimation of the long-term consequences of decisions concerning, for example, transportation, investments, and so forth, as the effects are passed along the various interaction loops of the system, and successive adjustments of the different agents occur. In general, we see that such a model offers a new basis for the understanding of "structure" resulting from the actions (choices) of the many agents of a system, having in part at least mutually dependent criteria of action (utility functions).

Concluding Remarks

The examples studied in the last section are quite removed from the simple systems of classical and quantum mechanics. However, it should be noted that there are no limits to structural stability. Every system may

FIGURE 5.17

A possible "history" of the urbanization of an initially uniform region, in which four economic functions seek to develop at each point in a network of fifty localities; the various attempts follow each other in a stochastic temporal sequence.

A. The distribution of population at time $t = 4$, on a lattice of 50 points. At $t = 0$, each has a population of 67.

B. At $t = 12$, the basic urban structure of the region is emerging, with five fast-growing centers.

C. By time $t = 20$, the structure has solidified, and the largest center exhibits the "urban sprawl" of residential suburbs.

D. At $t = 34$, growth of the urban centers is slow and the "above average growth" is taking place in the interurban zones, resulting in counterurbanization.

E. Evolution of the populations of points a, b, and c, indicated in part D, throughout the simulation.

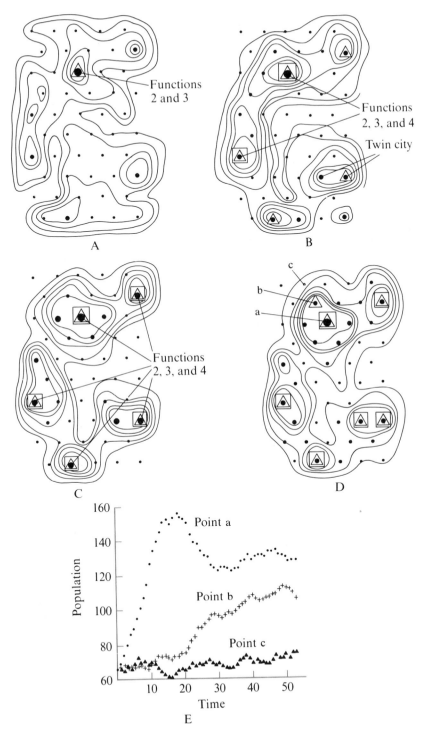

Functions
2 and 3

A

Functions
2, 3, and 4

Twin city

B

Functions
2, 3, and 4

C

b
c
a

D

E

present instabilities when suitable perturbations are introduced. There-fore, there can be no end to history. Ramon Margalef has, in a beautiful presentation, described what he calls the " baroque of the natural world " (Margalef 1976). He means that ecosystems contain many more species than would be " necessary " if biological efficiency alone were an organiz-ing principle. This " over creativity " of nature emerges naturally from the type of description being suggested here, in which " mutations " and " in-novations " occur stochastically and are integrated into the system by the deterministic relations prevailing at the moment. Thus, we have in this perspective the constant generation of " new types " and " new ideas " that may be incorporated into the structure of the system, causing its contin-ual evolution.

Jade pommel from China's Han period. Its diameter
is 4 cm. Private collection. Photograph by
R. Kayaert, Brussels.

Chapter 6

NONEQUILIBRIUM FLUCTUATIONS

The Breakdown of the Law
of Large Numbers

One reason why quantum mechanics has attracted such a great interest is
certainly the introduction of a probabilistic element into the description
of the microworld. As was seen in Chapter 3, in quantum mechanics
physical quantities are represented by operators that do not necessarily
commute. This leads to the well-known Heisenberg uncertainty relations.
Many people have seen in these relations a proof that on the microscopic
level, to which quantum mechanics applies, determinism is violated—a
statement that needs clarification.

As emphasized in Chapter 3, in the section on time change in quantum
mechanics, the basic equation of quantum mechanics, the Schrödinger
equation, is as deterministic as the classical equations of motion. There is
no uncertainty relation involving time and energy, in the sense in which
Heisenberg's uncertainty relations are valid. Once the wave function is

known at the initial time, we can, according to quantum mechanics, calculate its value at all times both in the past and in the future. Yet it is true that quantum mechanics introduces a basic probabilistic element in the description of the microworld. However, the macroscopic, thermodynamic description deals generally with averages, and the probabilistic elements introduced by quantum mechanics play no role. It is therefore of special interest to note that independent of the uncertainty relations there are *macroscopic* systems in which fluctuations and probabilistic description play an essential role. This can be expected in the neighborhood of bifurcations where the system has to "choose" one of the possible branches that appear at the bifurcation point. This statistical element will be analyzed in detail in this chapter in order to show that near bifurcations the *law of large numbers* essentially breaks down. (For an introduction to probability theory, see Feller, 1957.)

In general, fluctuations play a minor role in macroscopic physics, appearing only as small corrections that may be neglected if the system is sufficiently large. However, near bifurcations they play a critical role because there the fluctuation drives the average. This is the very meaning of the concept of *order through fluctuations*, which was introduced in Chapter 4.

It is interesting to note that this leads to unexpected aspects of chemical kinetics. Chemical kinetics is a field that is now about a hundred years old. It has always been formulated in terms of the type of rate equations, which were studied in Chapters 4 and 5. Their physical interpretation is quite simple. Thermal motion causes particles to collide. Most such collisions are elastic: that is, they change the translational kinetic energy (as well as the rotational and vibrational energy if polyatomic molecules are considered), without affecting the electronic structure. However, a fraction of these collisions are *reactive* and give rise to new chemical species.

On the basis of this physical picture, one may expect that the total number of collisions between two species of molecules, say X and Y, will be proportional to their concentrations, as will be the number of inelastic collisions. This idea has dominated the development of chemical kinetics since its formulation. However, how can such chaotic behavior, like that depicted by collisions occurring at random, ever give rise to coherent structure? Naturally some new feature must be taken into consideration; that is, the fact that, near instabilities, the distribution of reactive particles is no longer random. Until recently this feature was not included in chemical kinetics; however, further progress in its development is expected to take place in the next few years.

Before addressing the breakdown of the law of large numbers, let us briefly examine what is meant by this law. Consider, for example, a probability distribution of great importance in many fields of science and technology, the *Poisson distribution*. Suppose that we have a variable X that may take integral values, $X = 1, 2, 3, \ldots$. According to the Poisson distribution, the probability of X is given by

$$\mathrm{pr}(X) = e^{-\langle X \rangle} \frac{\langle X \rangle^X}{X!} \qquad (6.1)$$

This law is found to be valid in a variety of situations, such as the distribution of telephone calls, waiting time in restaurants, or the fluctuation of particles in a medium of given concentration. In equation 6.1, $\langle X \rangle$ represents the average value of X.

An important feature of the Poisson distribution is that $\langle X \rangle$ is the only parameter included in the distribution. The probability distribution is entirely determined by its mean. This is not so for the *Gaussian distribution* (equation 6.2), which contains in addition to the average, $\langle X \rangle$, the dispersion σ,

$$\mathrm{pr}(X) \sim e^{-(X - \langle X \rangle)^2/\sigma} \qquad (6.2)$$

From the probability distribution function, one can easily obtain the "variance," which gives the dispersion about the mean:

$$\langle \delta X^2 \rangle = \langle (X - \langle X \rangle)^2 \rangle \qquad (6.3)$$

The characteristic feature of the Poisson distribution is that the dispersion is equal to the average itself:

$$\langle \delta X^2 \rangle = \langle X \rangle \qquad (6.4)$$

Let us consider a situation in which X is an extensive quantity proportional to the number of particles N (in a given volume) or to the volume

V itself. We then obtain for the *relative* fluctuations the well-known square root law:

$$\frac{\sqrt{\langle \delta X^2 \rangle}}{\langle X \rangle} = \frac{1}{\sqrt{\langle X \rangle}} \sim \frac{1}{\sqrt{N}} \quad \text{or} \quad \frac{1}{\sqrt{V}} \tag{6.5}$$

The order of magnitude of the relative fluctuation is inversely proportional to the square root of the average. Therefore, for extensive variables of order N we obtain relative deviations of order $N^{-1/2}$. This is the characteristic feature of the law of large numbers. As a result we may disregard fluctuations in large systems and use a macroscopic description.

For other probability distributions, the mean square deviation is no longer equal to the average, as in equation 6.4. But whenever the law of large numbers applies, the order of magnitude of the mean square deviation is still the same, and we have

$$\frac{\langle \delta X^2 \rangle}{V} \sim \text{finite for } V \to \infty \tag{6.6}$$

We may also introduce a variable x into equation 6.2, which is "intensive"; that is, it does not increase with the size of the system (such as pressure, concentration, or temperature). The Gaussian distribution for such an intensive variable becomes, taking into account equation 6.6,

$$\text{pr}(x) \sim e^{-V(x - \langle x \rangle)^2 / \sigma} \tag{6.7}$$

This shows that the most probable deviation of an intensive variable from its mean will be of the order of $V^{-1/2}$ and will therefore become small when the system is large. Inversely, large fluctuations of intensive variables can occur only in small systems.

These remarks will be illustrated by examples to be considered later. We shall see how, near a bifurcation point, nature always finds some clever way to avoid the consequences of the law of large numbers through an appropriate nucleation process.

Chemical Games

To include fluctuations, we have to leave the macroscopic level. However, to turn to classical or quantum mechanics is practically out of the question. Every chemical reaction would then become an involved many-body problem. Therefore, it is useful to consider an intermediate level of somewhat the same type as that considered in Chapter 1 in the discussion of the random walk problem.

The basic idea is that of the existence of well-defined transition probabilities per unit time. Consider again the probability $W(k, t)$ of finding the Brownian particle at a place k, at time t. Let us introduce the transition probability $\omega_{\ell k}$, which gives us the probability (per unit time) for a transition between the two "states," k and ℓ. We may then express the time change of $W(k, t)$ in terms of a competition between *gain* terms, related to the transition $\ell \to k$, and *loss* terms, related to the transition $k \to \ell$. We can then obtain the basic equation

$$\frac{dW(k, t)}{dt} = \sum_{\ell \neq k} [\omega_{\ell k} W(\ell, t) - \omega_{k\ell} W(k, t)] \qquad (6.8)$$

In the Brownian motion problem, k would correspond to the position on the lattice *and* $\omega_{k\ell}$ would not be zero only if k differs from ℓ by one unit. But equation 6.8 is much more general. It is in fact the basic equation for Markov processes, which play a prominent role in modern theory of probability (see Barucha-Reid, 1960).

A characteristic feature of Markov processes is that the transition probabilities $\omega_{\ell k}$ involve only the states k and ℓ. The transition probability from $k \to \ell$ does not depend on which states were involved before the occupation of state k. In this sense the system has no memory. Markov processes have been used to describe many physical situations and can also be used to model chemical reactions. For example, let us consider a simple chain of unimolecular reactions:

$$A \underset{k_{21}}{\overset{k_{12}}{\rightleftharpoons}} X \underset{k_{32}}{\overset{k_{23}}{\rightleftharpoons}} E \tag{6.9}$$

The macroscopic kinetic equations are of the type introduced in Chapters 4 and 5 (we write here the kinetic constants):

$$\frac{dX}{dt} = (k_{12}A + k_{32}E) - (k_{21} + k_{23})X \tag{6.10}$$

We suppose, as before, that the concentrations of A and E are given. The steady state corresponding to equation 6.10 is

$$X_0 = \frac{k_{12}A + k_{32}E}{k_{21} + k_{23}} \tag{6.11}$$

In this standard macroscopic description, fluctuations are neglected. To study their effect, we introduce a probability distribution $W(A, X, E, t)$ and apply the general expression 6.8. The result is

$$\frac{dW(A, X, E, t)}{dt} = k_{12}(A + 1)W(A + 1, X - 1, E, t)$$

$$- k_{12}AW(A, X, E, t)$$

$$+ \text{ similar terms including } k_{21}, k_{23}, k_{32}$$

$$\tag{6.12}$$

The first term represents a *gain*. It corresponds to a transition from a state in which the number of particles A was $A + 1$ and the number of particles X was $X - 1$ to the state A, X by the decomposition of particle A occurring at the rate k_{12}. The second term represents, on the other hand, a *loss*. The state corresponded initially to the population A, X, E, but particle A has decomposed and we obtain the new state $A - 1, X + 1$. All other terms have a similar meaning.

This equation can be solved both for equilibrium and for nonequilibrium states. The result is a Poisson distribution with the macroscopic expression 6.11 as the mean for X.

This is quite satisfactory and seems so natural that for some time we believed that this result could be extended to all chemical reactions whatever their mechanism. But then a new, unexpected element entered. If we consider more general chemical reactions, the corresponding transition probabilities become *nonlinear*. For example, using the same argument as before, the transition probability corresponding to $A + X \rightarrow 2X$ is proportional to $(A + 1) \cdot (X - 1)$, the product of the number of particles of A and X before the inelastic collision. So the corresponding Markov equations also become *nonlinear*. It can be said that a distinct characteristic of chemical games is their nonlinearity as contrasted with the linear behavior of random walks, for which the transition probabilities are constant. To our surprise this new feature leads to *deviations* from the Poisson distribution. This unexpected result was proved by Gregoire Nicolis and myself (1971; see also Nicolis and Prigogine, 1977) and has aroused much interest. These deviations are very important from the point of view of the validity of macroscopic kinetic theory. We shall see that *macroscopic chemical equations are valid only when deviations from the Poisson distribution may be neglected.*

As an example, suppose that we have the chemical reaction $2X \rightarrow E$ with rate constant k. From the Markov equation 6.8, we may derive the time change of the average concentration of X. Not unexpectedly this leads to

$$\frac{d\langle X \rangle}{dt} = -k\langle X(X - 1) \rangle \tag{6.13}$$

Indeed, we have to choose two molecules in succession from the X molecules present. Note that

$$-\langle X(X - 1) \rangle = -\langle X \rangle^2 - (\langle \delta X^2 \rangle - \langle X \rangle) \tag{6.14}$$

in which we have used the identity $\langle \delta X^2 \rangle = \langle X^2 \rangle - \langle X \rangle^2$. The second term would vanish for a Poisson distribution, according to equation 6.4, which would mean that the behavior is governed by the macroscopic chemical equation.

This result is quite general. We see that deviations from the Poisson distribution play an essential role in the transition from microscopic to macroscopic level. Normally we may neglect them. For example, in expression 6.13, the first term must have the same order of magnitude as $\langle X \rangle$; that is, it must be proportional to the volume. The second is then independent of the volume. Therefore, in the limit of large volumes, it may be neglected. But, if the deviation from the Poisson behavior becomes proportional not to the volume itself as predicted by the law of large numbers, but to a higher power of the volume, then the whole macroscopic chemical description breaks down.

It is interesting to observe that in a sense chemical kinetics is very much a *mean field* theory, as are many other theories of classical physics and chemistry, such as the theory of equation of state (Van der Waal's theory), the theory of magnetism (Weiss field), and so forth. We also know from classical physics that such mean-field theories led to consistent results except near phase transitions. The theory initiated by Leo Kadanoff, Jack Swift, and Kenneth Wilson, among others, has as its basis the clever idea of studying long-range fluctuations that appear near critical points of phase transitions (see Stanley, 1971). The scale of the fluctuations becomes so large that molecular details no longer matter. The situation is rather similar here.

We would hope to find conditions ensuring the existence of *nonequilibrium* phase transitions for macroscopic systems by imposing length scale invariance on the master equation and taking the thermodynamic limit (i.e., the limit as both the number of particles and the volume tend to infinity, but density remains finite). From these conditions we should also be able to evaluate explicitly the way that the variance of fluctuations behaves near the transition. For nonequilibrium systems, this program has so far been carried out only for a simple model of master equation, namely, the Fokker-Planck equation (Dewel, Walgraef, and Borckmans 1977). Further work is in progress.

Let us now consider in detail a simple example in which the law of large numbers is violated.

Nonequilibrium Phase Transitions

Friedrich Schlögl has studied the following chemical sequence (see Schlögl, 1971, 1972; Nicolis and Prigogine, 1978):

$$A + 2X \underset{k_2}{\overset{k_1}{\rightleftharpoons}} 3X$$

$$X \underset{k_4}{\overset{k_3}{\rightleftharpoons}} B \tag{6.15}$$

Following our usual prescriptions, we can easily obtain the macroscopic kinetic equation

$$\frac{dX}{dt} = -k_2 X^3 + k_1 A X^2 - k_3 X + k_4 B \tag{6.16}$$

Suitable scaling and the introduction of the notations

$$\frac{X}{A} = 1 + x$$

$$\frac{B}{A} = 1 + \delta'$$

$$k_3 = 3 + \delta \tag{6.17}$$

reduces equation 6.16 to

$$\frac{dx}{dt} = -x^3 - \delta x + (\delta' - \delta) \tag{6.18}$$

The steady state is then given by the third-order algebraic equation

$$x^3 + \delta x = \delta' - \delta \tag{6.19}$$

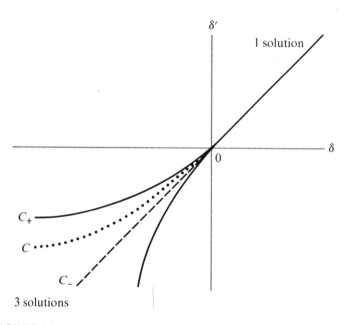

FIGURE 6.1
Behavior of the solutions of equation 6.19 in terms of the parameters δ and δ'; C represents the line of coexistence of multiple steady states.

It is interesting that this third-order equation is isomorphic with the one familiar from equilibrium phase transitions described by the Van der Waals theory. When we follow the evolution of the system along the line $\delta = \delta'$ (see Figure 6.1), we see that equation 6.19 has only one root, $x = 0$, for δ positive, whereas there are three roots, $x = 0$, $x_\pm = \pm\sqrt{-\delta}$, for δ negative (remember that x, being a concentration, must be real). This model is sufficiently simple to obtain the exact mean square deviation (see Nicolis and Turner, 1977a and b). As δ approaches zero, we obtain

$$\frac{\langle \delta X^2 \rangle}{V} \sim \frac{1}{\delta} \qquad (6.20)$$

Both these quantities tend to infinity as δ approaches zero, indicating a breakdown of the law of large numbers in the sense defined by equation 6.6. This breakdown becomes especially evident at the points at which the system can jump from the root x_+ to the root x_-, just as in an ordinary phase transition when a liquid phase becomes a vapor phase. At this point the variance is of the order of V^2. That is,

$$\frac{\langle \delta X^2 \rangle}{V^2} \sim \text{finite as } V \to \infty \qquad (6.21)$$

In other words, near nonequilibrium phase transitions there is no longer a consistent macroscopic description. Fluctuations are as important as average values.

One can show that in the multiple steady state region, the probability function $P(x)$ itself undergoes an extreme change in the limit $V \to \infty$. For any finite V, $P(x)$ is a double-humped distribution with peaks centered on the macroscopic stable states x_+ and x_-. For $V \to \infty$, each of the two humps collapses to a delta function (Nicolis and Turner, 1977a and b). Therefore a stationary probability is obtained of the form

$$P(x) = C_+ \delta(x - x_+) + C_- \delta(x - x_-) \qquad (6.22)$$

in which x is the intensive variable related to X, $x = X/V$. The weights C_+ and C_- sum to unity, and are otherwise determined explicitly from the master equation. Both $\delta(x - x_+)$ and $\delta(x - x_-)$ satisfy the master equation independently for $V \to \infty$. Their "mixture" (equation 6.22), on the other hand, gives the thermodynamic limit of the steady-state probability distribution evaluated first for finite system size. The analogy with equilibrium phase transitions of the Ising model type is striking: if x_+ and x_- were the values of total magnetization, then equation 6.22 would describe an Ising magnet at the zero (equilibrium) magnetization state. On the other hand, the "pure states," $\delta(x - x_+)$ and $\delta(x - x_-)$, would describe two magnetized states sustained for arbitrarily long times if appropriate boundary conditions are applied on the surface of the system.

The conclusion is not so astonishing as it at first seems. The very concept of macroscopic values, in a sense, loses its meaning. Macroscopic values are generally identified with the "most probable" values, which, if fluctuations may be neglected, become identical with average values. Here, however, we have near phase transition two most "probable" values, neither of which corresponds to the average value, and fluctuations between these two "macroscopic" values become very important.

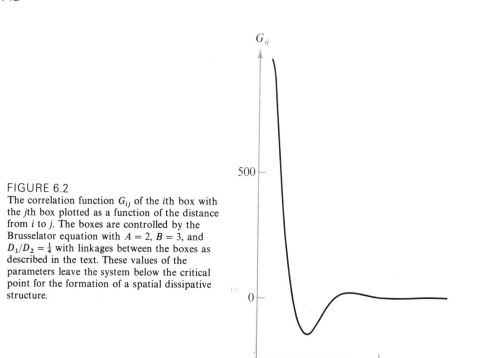

G_{ij}

500

0

0

10

The correlation function G_{ij} of the ith box with the jth box plotted as a function of the distance from i to j. The boxes are controlled by the Brusselator equation with $A = 2$, $B = 3$, and $D_1/D_2 = \frac{1}{4}$ with linkages between the boxes as described in the text. These values of the parameters leave the system below the critical point for the formation of a spatial dissipative structure.

Critical Fluctuations
in Nonequilibrium Systems

In equilibrium phase transitions, fluctuations near the critical point not only have a large amplitude, but also extend over *large distances*. Hervé Lemarchand and Gregoire Nicolis (1976) investigated the same problem for nonequilibrium phase transitions. To make the calculations possible, they considered a sequence of boxes. In each box the Brusselator type of reaction (reaction 4.57) takes place. In addition, there is diffusion between adjacent boxes. They calculated the correlation between the occupation numbers of X in two different boxes. One would expect that chemical inelastic collisions together with diffusion would lead to a chaotic behavior. But that is not so. Figures 6.2 and 6.3 show the correlation function for systems below the critical state and near it. It can be clearly seen that, near the critical point, chemical correlations are long range. The system acts as a *whole* in spite of the short-range character of the

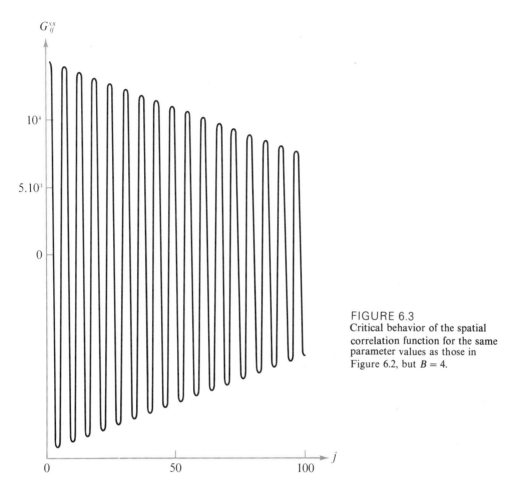

FIGURE 6.3
Critical behavior of the spatial correlation function for the same parameter values as those in Figure 6.2, but $B = 4$.

chemical interactions. Chaos gives rise to order.

What is the role of the number of particles in this process? That is an essential question to be addressed next, using the example of chemical oscillations.

Oscillations and Time Symmetry Breaking

The preceding considerations can also be applied to the problem of oscillating chemical reactions. From the molecular point of view, the existence of oscillations is very unexpected.

One might first think that it would be easier to obtain a coherent oscillating process with a few particles, say 50, than with as many as, say, Avogadro's number, 10^{23}, which are generally involved in macroscopic experiments. But computer experiments show that it is just the opposite. It is only in the limit of number of particles $N \to \infty$ that we tend to "long range" temporal order.

To understand this result at least qualitatively, let us consider the analogy with phase transitions. If we cool down a paramagnetic substance to a temperature called the Curie point, the system undergoes a behavioral change and becomes ferromagnetic. Above the Curie point, all directions play the same role; below it, there is a privileged direction corresponding to the direction of magnetization.

Nothing in the macroscopic equation determines which direction the magnetization will take. In principle, all directions are equally likely. If the ferromagnet contained a *finite* number of particles, this privileged direction would not be maintained in time. It would rotate. However, in an infinite system, no fluctuation whatsoever can shift the direction of the ferromagnet. The long-range order is established once and for all.

The situation is very similar in oscillating chemical reactions. It can be shown that, when the system switches to a limit cycle, the stationary probability distribution also undergoes a structural change: it switches from a single-humped form to a craterlike surface centered on the limit cycle. As in equation 6.22, the crater gets sharper as V increases and, in the limit $V \to \infty$, it becomes singular. In addition to this, however, a family of *time-dependent* solutions of the master equation appears. For any finite V, these solutions lead to damped oscillations so that the only long-time solution remains the steady-state one. Intuitively, this means that the *phase* of the motion on the limit cycle, which plays the same role as the direction of magnetization, is determined by the initial conditions. If the system is finite, fluctuations will progressively take over and destroy phase coherence.

On the other hand, computer simulations suggest that as V increases the time-dependent modes are less and less damped. We can therefore expect to obtain, in the limit $V \to \infty$, a whole family of time-dependent solutions of the master equation rotating along the limit cycle (Nicolis and Malek-Mansour 1978). Again, in our intuitive picture, this would mean that in an infinite system phase coherence can be maintained for arbitrarily long times, just as a privileged initial magnetization can be sustained in a ferromagnet. In this sense, therefore, the appearance of a periodic reaction is a *time symmetry breaking process*, exactly as ferromagnetism is a space symmetry breaking one.

The same observations could be made for time-independent but space-dependent dissipative structures. In other words, it is only if the chemical equations are exactly valid (i.e., in the limit of large numbers when the law of large numbers applies) that we may have coherent nonequilibrium structures.

An additional element to the far-from-equilibrium condition employed in Chapter 4 is the size of the system. If life is indeed associated with coherent structures—and everything supports this view—it must be a *macroscopic phenomenon* based on the interaction of a large number of degrees of freedom. It is true that some molecules, such as nucleic acids, play a dominant role, but they can only be generated in a coherent medium involving a large number of degrees of freedom.

Limits to Complexity

The methods outlined in this chapter may be applied to many situations. One of the interesting features of this approach is that it shows that the laws of fluctuation depend markedly on the scale. The situation becomes quite similar to that in the classical theory of the nucleation of a liquid drop in a supersaturated vapor. A droplet smaller than a critical size (called the size of an "embryo") is unstable, whereas if it is larger than this size it grows and changes the vapor into a liquid (see Figure 6.4).

Such a nucleation effect also appears in the formation of an arbitrary dissipative structure (see Nicolis and Prigogine, 1977). We may write a master equation of the type

$$\frac{\partial P_{\Delta V}}{\partial t} = \text{chemical effects inside } \Delta V$$

$$+ \text{diffusion with the outside world} \tag{6.23}$$

which takes into account *both* the effect of the chemical reaction inside a volume ΔV *and* the migration of the particles through exchange with the outside world. The form of this equation is very simple. When the average

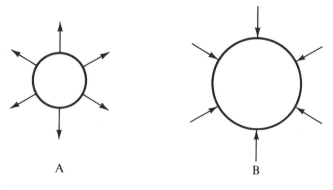

FIGURE 6.4
Nucleation of a liquid droplet in supersaturated vapor: (A) a droplet smaller than the critical size; (B) a droplet larger than the critical size.

$\langle X^2 \rangle$ in volume ΔV is calculated, one obtains from equation 6.2 the sum of two terms represented schematically as

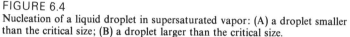

$$\frac{d\langle X^2 \rangle_{\Delta V}}{dt} = \text{chemical effects inside } \Delta V$$

$$- 2\mathscr{D}[\langle \delta X^2 \rangle_{\Delta V} - \langle X \rangle_{\Delta V}] \qquad (6.24)$$

The first term is the effect of the chemistry inside the volume ΔV. The second is due to the exchange with the outside world. Coefficient $\mathscr{D}$ increases when the surface-to-volume ratio becomes large. The interesting point is that the second term contains the exact difference between the mean square fluctuation and the average. For sufficiently small systems, this will be the dominant contribution, and the distribution will become Poissonian in accordance with equation 6.4. In other words, the outside world always acts as a mean field that tends to *damp* the fluctuations through the interactions taking place on the boundaries of the fluctuating region. This is a very general result. For small-scale fluctuations, boundary effects will dominate and fluctuations will regress. However, for large-scale fluctuations, boundary effects become negligible. Between these limiting cases lies the actual size of nucleation.

This result is of interest for a very general question long discussed by ecologists: the problem of the *limits to complexity* (May 1974). Let us for a moment return to the linear stability analysis that was developed in Chapter 4. This leads to some dispersion equation. The degree of this equation is equal to the number of interacting species. Therefore, in a complex medium, such as a tropical forest or a modern civilization, the degree of such an equation would be very high indeed. Consequently, the chances of having at least one positive root leading to instability increase. How then is it possible that complex systems exist at all? I believe that the theory summarized here gives the beginning of an answer. The coefficient in equation 6.24 measures the degree of coupling between the system and its surroundings. We may expect that in systems that are very complex, in the sense that there are many interacting species or components, this coefficient will be large, as will be the size of the fluctuation, which could start the instability. Therefore we reach the conclusion that a sufficiently complex system is generally in a *metastable* state. The value of the threshold depends both on the system's parameters and on the external conditions. The limits to complexity is not a one-sided problem. It is interesting to note that, in recent numerical simulations of nucleation, this role of communication (e.g., through diffusion in nucleation) has been implemented.

Effect of Environmental Noise

So far we have been concerned with the dynamics of *internal fluctuations*. We have seen that these fluctuations, which are generated spontaneously by the system itself, tend to be small except when the system is near a bifurcation or in the coexistence region of simultaneously stable states.

On the other hand, the parameters of a macroscopic system—including most of the bifurcation parameters—are externally controlled quantities and are therefore also subject to fluctuations. In many cases, the system's environment fluctuates violently. It can be expected, therefore, that such fluctuations, which are perceived by the system as an "external noise," could deeply affect its behavior. This point was established recently both theoretically (Horsthemke and Malek-Mansour 1976; Arnold, Horsthemke, and Lefever 1978; Nicolis and Benrubi 1976) and experimentally (Kawakubo, Kabashima, and Tsuchiya 1978). It seems that environmental fluctuations can both *affect bifurcation* and—more spectacularly—*generate new nonequilibrium transitions* not predicted by the phenomenological laws of evolution.

The traditional approach to environmental fluctuations originated with Paul Langevin's analysis of the Brownian motion problem. In this view, the rate function [say $v(x)$] describing the macroscopic evolution of an observable quantity (say x) gives only part of the instantaneous rate of change of x. Because of fluctuations of the surroundings, the system also experiences a random force $\mathbf{F}(x, t)$. Thus, considering x to be a fluctuating quantity, we write

$$\frac{dx}{dt} = v(x) + \mathbf{F}(x, t) \qquad (6.25)$$

If, as in Brownian motion, $\mathbf{F}$ reflects the effect of intermolecular interactions, its successive values must be uncorrelated both in time and in space. Because of this, the variance of fluctuations obtained agrees with the central limit theorem. On the other hand, in a nonequilibrium environment, fluctuations can modify the macroscopic behavior of the system dramatically. It seems that, for this behavior to occur, the external noise must act *multiplicatively* rather than additively; that is, it is coupled with a function of the state variable of x, which vanishes if x itself vanishes.

To illustrate this point, consider a modified Schlögl model (see equation 6.15):

$$A + 2X \rightleftharpoons 3X$$

$$B + 2X \rightarrow C$$

$$X \rightarrow D \qquad (6.26)$$

We set all rate constants equal to one and

$$\gamma = A - 2B \qquad (6.27)$$

The phenomenological equation is

$$\frac{dx}{dt} = -x^3 + \gamma x^2 - x \qquad (6.28)$$

At $\gamma = 2$, both a stable steady-state solution and an unstable one emerge, as shown in Figure 6.5. In addition, $x = 0$ is always a solution that is

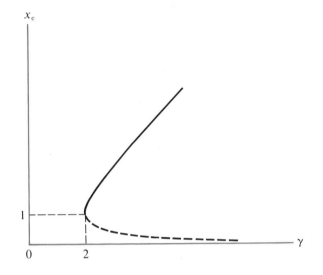

FIGURE 6.5
Stationary solutions x_0 of equation 6.28 versus γ: solid curve indicates
stable solution; dashed curve, unstable.

stable under infinitesimal perturbations.

We now consider γ to be a random variable. The simplest assumption
is that it corresponds to a *Gaussian white noise*, just as in Brownian
motion problems. We set

$$\langle \gamma \rangle = P$$

$$\langle \gamma^2 \rangle = \sigma^2 \tag{6.29}$$

Instead of writing equation 6.28, we now write a *stochastic differential
equation* (Arnold 1973), which is a suitable generalization of the Langevin
equation (equation 6.25), free of some ambiguities inherent in the usual
formulation of the latter. This equation couples the noise with the second
power, x^2, of the state variable. It can be connected to a master equation
of the Fokker-Planck type, from which the stationary probability distribution can be computed. The result is that in this distribution the transition point $\gamma = 2$ of the phenomenological description disappears: the
process certainly reaches zero and subsequently remains there.

In the experimental work on the effects of noise mentioned near the
beginning of this section (Kawakubo, Kabashima, and Tsuchiya 1978)
the arrangement is very similar to that expressed in equation 6.28, except

that noise is coupled with a linear term and equation 6.28 includes a constant input term. As it turns out, for small values of the variance σ^2 the system (a parametric oscillator circuit) exhibits limit-cycle behavior. However, if the variance exceeds a threshold, the oscillatory behavior disappears and the system falls to a steady-state regime.

Concluding Remarks

We have now outlined the main elements of the physics of becoming. Many unexpected results have been reported, extending the range of thermodynamics. Classical thermodynamics was associated, as mentioned, with the forgetting of initial conditions and the destruction of structures. We have seen, however, that there is another *macroscopic* region in which, within the framework of thermodynamics, structures may spontaneously appear.

The role of determinism in macroscopic physics must be reappraised. Near instabilities, there are large fluctuations that lead to a breakdown of the usual laws of probability theory. A new view of chemical kinetics has emerged. As a consequence of these developments, classical chemical kinetics appears as a *mean field* theory, but to describe the appearance of coherent structures, to describe the formation of order from chaos, we must introduce a new, more refined description of the temporal sequences that lead to the time evolution of the system. However, the stabilization of dissipative structures requires a large number of degrees of freedom. This is the reason that a deterministic description prevails between successive bifurcations.

Both the physics of being and the physics of becoming have taken new dimensions in the past several years. Can the two points of view be unified in some way? After all, we are living in a single world whose aspects, however diverse they seem at first, must have some relation. This is the subject of Part IV.

Part III

THE BRIDGE
FROM BEING TO BECOMING

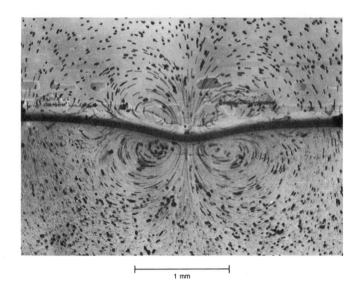

H——————————H
1 mm

Flow Lines in Roll Cells.

THE GROWTH OF ROLL CELLS

Roll cells develop at the interface between isoamylol and water in the presence of the surfactant sodium hexadecyl sulfate, a common detergent. The instability, called a Margangoni instability, is caused by variations in the surface tension depending on the concentration of the surfactant. This instability and the coupling between diffusion and convection lead to formation of the roll cells shown on the facing page. The width of the roll cells increases roughly as the square root of elapsed time, which is to be expected for a process governed by diffusion.

The figures on the facing page are schlieren photographs taken at the times indicated. The final photograph, taken 600 seconds after the experiment began, shows both the large secondary roll cells and the small primary roll cells (see the diagram below the schlierin photographs). In all these pictures and in the diagram, the upper phase, phase I, was a 1-percent solution of hexadecyl sulfate in isoamylol and the lower phase, phase II, was pure water initially. The composition is changed by active transport during the experiment.

The photograph above shows the flow lines made visible by adding small particles of aluminum to the liquids. This photograph was taken 15 seconds after the start of the experiment with an exposure time of one-fifth of a second.

Photographs by H. Linde, P. Shwartz, and H. Wilke. Reproduced with permission.

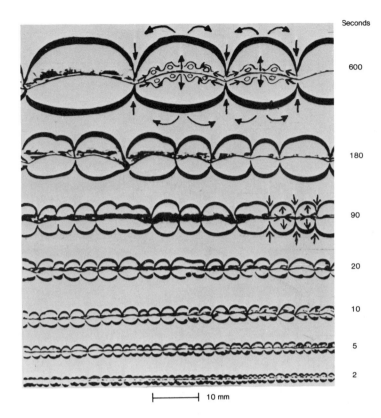

Seconds

600

180

90

20

10

5

2

|——| 10 mm

Schlieren Photographs of Roll Cells. The cells grow larger as the concentration gradient is decreased by active transport across the capillary split.

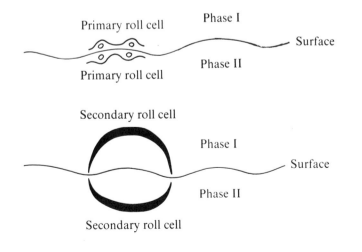

Primary roll cell   Phase I

Surface

Phase II
Primary roll cell

Secondary roll cell

Phase I

Surface

Phase II

Secondary roll cell

Chapter 7

KINETIC THEORY

Introduction

The relation between the two basic fields of theoretical physics—dynamics and thermodynamics—is probably the most challenging problem to be treated in this book. It has been a subject of discussion since the formulation of thermodynamics one hundred fifty years ago, and thousands of papers have been written about it. The relation touches upon the meaning of time and is, therefore, of crucial importance. We cannot expect an easy solution to the problem because, if that were possible, it would have been solved long ago. I shall give qualitative arguments to justify my conviction that we have now found a way of avoiding what seemed for so long to have been insurmountable obstacles. However, because no "proofs" are given here, the interested reader should consult a monograph in preparation on this subject (Prigogine, forthcoming).*

---

* More details may also be found in the appendixes to this book.

We shall start with kinetic theory, and especially with Boltzmann's *H* theorem, which must be considered a milestone on the road to understanding the microscopic meaning of entropy (for a presentation of classical kinetic theory, see Chapman and Cowling, 1970).

Why was Boltzmann so fascinated with the second law? What attracted him to such an extent that he devoted virtually his entire career to understanding and interpreting it? In *Populäre Schriften* (1905), he wrote: "If one would ask me which name we should give to this century, I would answer without hesitation that this is the century of Darwin." Boltzmann was deeply attracted by the idea of evolution, and his ambition was to become the "Darwin" of the evolution of matter.

Boltzmann's approach had astounding successes. It has left a deep imprint on the history of physics. The discovery of the quantum by Planck was an outcome of Boltzmann's approach. I fully share the enthusiasm with which Schrödinger wrote in 1929 that "His [Boltzmann's] line of thought may be called my first love in science. No other has ever thus enraptured me or will ever do so again." Yet, it must be recognized that there are serious difficulties with Boltzmann's approach. It proved very difficult to apply his approach except to gases at low concentration. Although modern kinetic theory has been quite successful in discussing some aspects of transport theory involving viscosity, heat conductivity, and so forth, it sheds no light on the microscopic meaning of entropy in dense systems. Even for gases at low concentration, as we shall see, Boltzmann's definition of entropy applies only for certain initial conditions.

It is because of such difficulties that Gibbs and Einstein worked out a much more general approach in terms of ensemble theory, which was described in Chapters 2 and 3. Their approach was, however, essentially limited to systems in equilibrium. The complete title of Gibbs's classic memoir is *Elementary Principles of Statistical Mechanics: Development with Special Reference to the Rational Foundations of Thermodynamics* (Gibbs 1902). This work on (equilibrium) thermodynamics is far from Boltzmann's ambition to derive a mechanical theory of the evolution of matter. Because of the lack of success in attempts to apply ensemble theory to nonequilibrium situations (see the sections on Gibbs's entropy and the Poincaré-Misra theorem later in this chapter), the idea that supplementary approximations must be introduced to deal with nonequilibrium became prevalent. Gibbs's well-known example of the mixture of ink with water was mentioned in Chapter 1. However, this idea of supplementary "coarse graining" has not been successful (though it appealed to many physicists) because it proved in the end to be as difficult to provide a precise prescription for coarse graining as it is to solve the problem of the microscopic meaning of irreversibility itself.

Today we understand the nature of these difficulties a little better and as a result we may follow a path designed to avoid them. First, it should be emphasized that Boltzmann's approach goes beyond dynamics; it uses a remarkable mixture of dynamical and probabilistic concepts. In fact Boltzmann's kinetic equation is the forerunner of the Markov chains that were used to model chemical equations in Chapter 6.

In his *Leçons de thermodynamique*, Poincaré discussed in detail the relation of the second law with classical dynamics. Yet he did not even quote Boltzmann! Moreover, his conclusion is categorical: thermodynamics and dynamics are *incompatible*. He based his conclusion on a short paper he had published earlier (1889) in which he proved that, in the framework of Hamiltonian dynamics, there can be no function of coordinates and momenta that would have the properties of a Lyapounov function (see the section titled The Poincaré-Misra Theorem later in this chapter and that titled The Second Law of Thermodynamics in Chapter 1).

As Misra has shown recently, this conclusion remains valid even in the framework of ensemble theory. The importance of the Poincaré-Misra theorem is that it leaves us only two alternatives. We can conclude with Poincaré that there is no dynamical interpretation of the second law. Then, irreversibility comes from supplementary phenomenological or subjectivistic assumptions, from "mistakes." But how then can we account for the wealth of important results and concepts that derive from the second law?* In a sense living beings, we ourselves, are then "mistakes."

Fortunately, there is a second alternative. Poincaré tried to associate entropy with a function of correlations and momenta, but this attempt also failed. Can we not retain the idea of introducing a microscopic entropy such that macroscopic entropy is an appropriate average of the microscopic entropy, thus realizing Poincaré's program in a different

---

* Refer to the discussions in Chapters 4 and 5 that stress the importance of dissipative structures for biological problems. How can we account for these results if the second law is an approximation?

way? Quantum mechanics has accustomed us to associate *operators* with physical quantities. Moreover, we have seen that in the ensemble approach (see the section dealing with ensemble theory in Chapter 2) the time evolution is described by the Liouville *operator*.* It therefore becomes very tempting to try to realize Poincaré's program in terms of an operator associated with the microscopic entropy (or Lyapounov functions).

At first this seems a strange idea—or at least a purely formal device. An attempt will be made here to show that this is not so, that the idea of introducing a microscopic entropy operator is on the contrary a very simple and natural one. It should be remembered that the idea of an energy operator (the Hamiltonian operator H_{op} referred to in Chapter 3) means that we cannot associate a well-defined value of energy with an arbitrary wave function *unless* it happens to be an eigenfunction of H_{op}. Similarly the idea of an entropy operator would mean that the relation between distribution function ρ and entropy would be more subtle than formerly considered. Again, in general we could not associate a well-defined value of entropy with the distribution function (or a function of ρ) unless it happens to be an eigenfunction of this operator.

As will be seen, this more-refined relation between density ρ and entropy is in line with the idea of randomness on the microscopic level as introduced in classical mechanics by the concept of weak stability (see Chapter 2). We can expect therefore that the construction of this operator will be possible only if the basic concepts of classical (or quantum) mechanics, such as trajectories or wave functions, correspond to unobservable idealizations. Whenever it is possible to introduce such a microscopic entropy operator, classical dynamics becomes an algebra of noncommuting operators (somewhat like quantum mechanics). It is certainly a great surprise that such a fundamental change in the structure of dynamics can be forced on us by the concept of irreversibility. Basically the same conclusions apply to quantum mechanics, of which the consequent fundamental change in structure will be described briefly in Chapter 8 and in Appendix C.

---

* We have already seen that the use of operators becomes natural whenever we give up the idea of trajectory (see also Appendixes A and B). Certainly the idea of operators is not restricted to quantum mechanics.

In short, the usual formulation of classical (or quantum) mechanics has become "embedded" in a larger theoretical structure, which also allows the description of irreversible processes. It is very gratifying that irreversibility does not correspond to some approximation added to the laws of dynamics but to an enlargement of their theoretical structure.

In this framework, there is a new type of complementarity between the dynamical description and entropy. It can be expected that this complementarity exists only if the dynamical system is sufficiently "complex." Nobody would expect a thermodynamic type of behavior for a simple harmonic oscillator.

In this chapter, Boltzmann's approach is discussed and the Poincaré-Misra theorem presented. The construction of a new form of classical or quantum dynamics that explicitly displays irreversible processes will be presented in Chapter 8.

Boltzmann's Kinetic Theory

A few years before the publication in 1872 of Boltzmann's fundamental paper, "Further Studies on Thermal Equilibrium between Gas Molecules," Maxwell had already studied the evolution of the velocity distribution function, $f(r, v, t)$, which gives the number of particles having at time t the position r and the velocity v (Maxwell 1867). (In terms of the general distribution function ρ, as defined in equation 2.8, f is obtained by integrating over all coordinates and momenta except those of a single molecule.) Maxwell gave convincing arguments that, for long times in low-concentration gases, this velocity distribution should tend to the Gaussian form,

$$f(r, v, t) \rightarrow \left(\frac{m}{2\pi kT}\right)^{3/2} e^{-mv^2/2kT} \qquad (7.1)$$

in which m is the mass of the molecules and T the (absolute) temperature (see equation 4.1). This is the well-known Maxwell velocity distribution. Boltzmann's aim was to discover a molecular mechanism that would ensure the validity of Maxwell's velocity distribution for long times. His starting point was to deal with large systems including many particles. He considered it natural that, like social and biological situations, such systems would call attention not to individual particles, but to the evolution of groups of particles, and concepts of probability could be used quite freely. He decomposed the time variations of the velocity distribution into two terms, one due to the motion of the particle, the other due to binary collisions:

$$\frac{\partial f}{\partial t} = \left(\frac{\partial f}{\partial t}\right)_{\text{flow}} + \left(\frac{\partial f}{\partial t}\right)_{\text{coll}} \qquad (7.2)$$

There is no difficulty in making the flow term *explicit*. We must simply introduce the Hamiltonian for free particles, $H = p^2/2m$, and apply equation 2.11. We then obtain

$$\left(\frac{\partial f}{\partial t}\right)_{\text{flow}} = -\frac{\partial H}{\partial p}\frac{\partial f}{\partial x} = -v\frac{\partial f}{\partial x} \qquad (7.3)$$

in which $v = p/m$ is the velocity. However, the evaluation of the collision term does present a problem. Boltzmann used a plausibility argument very similar to the type of arguments introduced in the theory of Markov chains, which were described in Chapter 5. Historically, however, Boltzmann's theory preceded the theory of Markov chains.

As was done in equation 6.8, Boltzmann decomposed the time change due to collision into a *gain* term, in which one particle with velocity v appears at point r (that means in some element of volume around point r), and *loss* terms, in which such a molecule disappears because of collisions. Therefore we have the scheme

$$v', v'_1 \rightarrow v, v_1 \quad \text{gain}$$

$$v, v_1 \rightarrow v', v'_1 \quad \text{loss} \qquad (7.4)$$

The frequency of these collisions is proportional to the number of molecules that have velocities v', v'_1 (or v, v_1); that is, $f(v')f(v'_1)$ [or $f(v)f(v_1)$]. After a few elementary calculations, this gives the contribution for the collision term (see Chapman and Cowling, 1970):

$$\left(\frac{\partial f}{\partial t}\right)_{coll} = \iint d\omega \, dv_1 \sigma[f'f'_1 - ff_1] \tag{7.5}$$

The integration is performed both on the geometrical factors that determine the collision cross section σ and over the velocity v_1 of one of the molecules in the collision. Adding equations 7.3 and 7.5, we obtain Boltzmann's celebrated integro-differential equation for the velocity distribution:

$$\frac{\partial f}{\partial t} + v \frac{\partial f}{\partial x} = \iint d\omega \, dv_1 \sigma[f'f'_1 - ff_1] \tag{7.6}$$

After this equation has been obtained, we can introduce Boltzmann's H-quantity:

$$H = \int dvf \log f \tag{7.7}$$

and prove that

$$\frac{\partial H}{\partial t} = -\iint d\omega \, dv_1 \sigma \log \frac{f'f'_1}{ff_1}(f'f'_1 - ff_1) \leqslant 0 \tag{7.8}$$

as a result of the simple inequality

$$\log \frac{a}{b} \cdot (a - b) \geqslant 0 \tag{7.9}$$

We therefore obtain a *Lyapounov function*. However, the basic difference between this Lyapounov function and that considered in Chapter 1 in the section on the second law of thermodynamics is that it is now expressed in terms of the velocity distribution and not in terms of macroscopic quantities such as temperature.

The Lyapounov function reaches its minimum when the condition

$$\log f + \log f_1 = \log f' + \log f'_1 \tag{7.10}$$

is satisfied. This condition has a simple meaning in terms of the *collisional invariants*, which are the number of particles, the three Cartesian momenta of the particles, and the kinetic energy. These five quantities are conserved in a collision. Therefore log f must be a linear expression of these given quantities, and disregarding the momenta, which are only important if there is motion as a whole, we immediately arrive at the Maxwell distribution (formula 7.1), in which, indeed, log f is a linear function of the kinetic energy $mv^2/2$.

Boltzmann's kinetic equation is a very complicated one because it contains the product of the unknown distribution functions under the integral. For systems near equilibrium, we may write

$$f = f^{(0)}(1 + \phi) \qquad (7.11)$$

in which $f^{(0)}$ is the Maxwell distribution and ϕ is considered a small quantity. We then obtain a linear equation for ϕ, which has proved to be extremely useful in transport theory. An even cruder approximation of Boltzmann's equation is to replace the whole collision term by a linear relaxation term, and to write

$$\frac{\partial f}{\partial t} + v \frac{\partial f}{\partial x} = - \frac{f - f^{(0)}}{\tau} \qquad (7.12)$$

in which τ is an average relaxation time that gives an order of magnitude of the time interval necessary to reach the Maxwell distribution.

Boltzmann's equation has given rise to many other kinetic equations that are valid under rather similar conditions (collisions between excitations in solids, plasmas, etc.). More recently, extensions to dense systems have been suggested. However, these generalized equations for dense media do not admit a Lyapounov function and the connection with the second law is lost.

The procedure for using Boltzmann's approach can be summarized as follows:

Dynamics

↓

Kinetic equation ("Markov process")

↓

Entropy (through H)

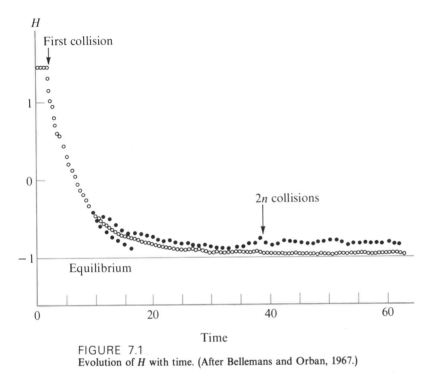

FIGURE 7.1
Evolution of H with time. (After Bellemans and Orban, 1967.)

In recent years there have been many numerical calculations to verify Boltzmann's predictions. The H-quantity has been calculated on computers, for example, for two-dimensional hard spheres (hard disks), starting with disks on lattice sites with isotropic velocity distribution (Bellemans and Orban 1967). The results, which are given in Figure 7.1, confirm Boltzmann's prediction.

Boltzmann's theory has also been used to calculate transport properties (viscosity and thermal conductivity). This is the great achievement of methods devised by Sydney Chapman and David Enskog for the solution of Boltzmann's equation. In this case, too, agreement has been quite satisfactory (see Chapman and Cowling, 1970, and Hirschfelder, Curtiss, and Bird, 1954).

Why does Boltzmann's method work? The first aspect to consider is the assumption of *molecular chaos*. As was discussed in Chapter 5 in the section on classical chemical kinetics, Boltzmann calculated the *average* number of collisions, neglecting fluctuations. But this is not the only important element. If we compare Boltzmann's equation with the Liouville equation (2.12), we see that in Boltzmann's equation the symmetry of the Liouville equation is broken. If we change $L \to -L$ and $t \to -t$ in the Liouville equation, this equation remains invariant. We can change $L \to -L$ by changing the momentum (or the velocity) $p \to -p$. This is a consequence of equation 2.13. By looking at Boltzmann's kinetic equation, or the simple version given in equation 7.12, we see that the flow term changes sign when v is replaced by $-v$, but the collision term remains invariant. This term is *even* with respect to velocity inversion. This is also true for the original Boltzmann equation.

Therefore the symmetry of the collision term violates the "$L - t$" symmetry of the Liouville equation. A characteristic feature of Boltzmann's equation is that it possesses a *new type of symmetry*, one that does not appear in the Liouville equation, neither in classical nor in quantum mechanics. In brief, the time evolution contains both odd and even terms in L.

This is very important. Only the collision term (which is even in L) contributes to the evolution of the Lyapounov function H. We may say that Boltzmann's equation transposes the basic thermodynamic distinction between reversible and irreversible processes into the microscopic (or more accurately kinetic) description. The flow term corresponds to a reversible process and the collision term to an irreversible process. Thus, there is a close correspondence between the thermodynamic description and Boltzmann's description, but unfortunately this correspondence is not "deduced" from dynamics; it is postulated from the start (i.e., equation 7.2).

A surprising feature of Boltzmann's theorem is its universal character. The interaction between the molecules may be quite varied: we may consider hard spheres, repulsive central forces decreasing according to some power law, or both repulsive and attractive forces. Yet, *independently* of the microscopic interactions, the H quantity has a universal form. We shall return to the interpretation of this remarkable feature in the next chapter. Let us turn now to some of the difficulties related to Boltzmann's treatment of kinetic theory.

Correlations and the Entropy
of Rejuvenation

It has already been mentioned that, in spite of their successes, Boltzmann's ideas have met with both practical and theoretical difficulties. For example, it seemed impossible to extend the construction of the H-quantity to other systems, such as dense gases or liquids. It is easy to see that the practical and the theoretical difficulties are related. Let us first concentrate on the theoretical difficulties. From the start, Boltzmann's ideas met with strong objections. Poincaré went so far as to write that he could not recommend the study of Boltzmann's paper because the premises in Boltzmann's considerations clashed with his conclusions (Poincaré 1893). (Poincaré's point of view will be returned to later in this chapter.)

Other objections were formulated in the form of paradoxes. One was Ernst Zermelo's recurrence paradox, which is based on the celebrated Poincaré theorem that states that, "for almost all initial states, an arbitrary function of phase space will infinitely often assume its initial value, within arbitrary error, provided the system remains in a finite part of the phase space. As a result it seems that irreversibility is incompatible with the validity of this theorem." (For references to the paradoxes, see Chandrasekhar, 1943.)

As has been pointed out recently, notably by Joel Lebowitz (see Rice, Freed, and Light, 1972), Zermelo's objection is not justified, because Boltzmann's theory deals with the distribution function f, whereas Poincaré's theorem refers to single trajectories.

We could then ask, Why introduce a distribution function formalism at all? At least for classical dynamics, we already know the answer (see Chapter 2). Whenever we have weak stability as in mixing systems or in dynamical systems exhibiting the Poincaré catastrophe, we cannot perform the transition from a statistical distribution function to a well-defined trajectory. (For quantum mechanics, see Appendix C.)

This is a very important point: for *any* dynamical system we never know the exact initial conditions and therefore the trajectory. Yet the transition from the distribution function in phase space to the trajectory corresponds to a well-defined process of successive approximations. However, for systems exhibiting "weak stability," there is no process of successive approximations, and the concept of a trajectory corresponds to an idealization beyond that which can be obtained from experiments, regardless of their accuracy.

Another serious objection is based on Joseph Loschmidt's reversibility paradox: because the laws of mechanics are symmetrical with respect to the inversion $t \rightarrow -t$, to each process there corresponds a time-reversed process. This also seems to be in contradiction with the existence of irreversible processes.

Is Loschmidt's paradox at all justified? It is easy to test it by means of a computer experiment. André Bellemans and John Orban (1967) have calculated Boltzmann's H-quantity for two-dimensional hard spheres (hard disks). They start with disks on lattice sites with an isotropic velocity distribution. The results are shown in Figure 7.2.

We see that, indeed, the entropy (that is, minus H) first decreases after the velocity inversion. The system deviates from equilibrium over a period ranging from fifty to sixty collisions (which would correspond in a low-concentration gas to about 10^{-6} seconds).

The situation is similar for spin-echo experiments and plasma-echo experiments. Over limited periods, anti-Boltzmannian behavior in this sense can be observed. All this shows that Boltzmann's equation is not always applicable. Paul and Tatiana Ehrenfest made the remark that Boltzmann's equation cannot be correct *both* before *and* after inversion of velocities (see Ehrenfest and Ehrenfest, 1911).

Boltzmann's view was that, in some sense, physical situations for which the kinetic equation (equation 7.6) is valid would be overwhelmingly more frequent. It is difficult to accept this view, because today we can realize both computer and laboratory experiments in which his kinetic equation is *not valid*, at least over limited periods.

What inference can be drawn from the fact that there are situations for which the kinetic equation is valid and others for which it is not? Does this fact express a limitation of Boltzmann's statistical interpretation of entropy or a failure of the second law for some class of initial conditions?

The physical situation is quite clear: velocity inversion creates correla-

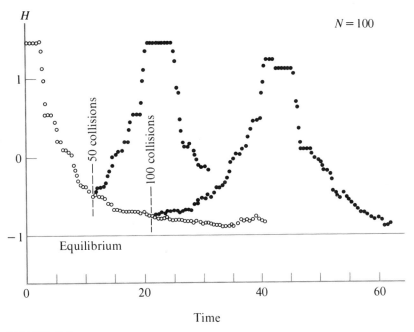

FIGURE 7.2
Evolution of H with time for a system of 100 disks when velocities are inverted after 50 collisions (open circles), 100 collisions (solid circles). (After Bellemans and Orban, 1967.)

tions between particles that may be of macroscopic range.* Particles that collide at time t_1 must collide again at time $2t_0 - t_1$. These anomalous correlations can be expected to disappear during the period from t_0 to $2t_0$, after which the system returns to a "normal" behavior.

In brief, entropy production can be understood in the interval 0 to t_0 to be associated with the "Maxwellianization" of the velocity distribution, whereas in the period t_0 to $2t_0$ it should be associated with the decay of anomalous correlations.

Thus the failure of Boltzmann's approach to cope with such situations can be easily understood. We need a statistical expression of entropy that depends *explicitly* on correlations. Let us briefly consider how the H-quantity would evolve if we were able to construct a Lyapounov function that also contained the correlations (see Prigogine et al., 1973).

---

* These "anomalous" correlations also have the property that they exist prior to collisions, whereas the normal correlations are produced by the collisions.

For example, consider the positive quantity

$$\Omega = \int \rho^2 \, dp \, dq > 0 \qquad (7.13)$$

in which we integrate over the phase space. In quantum mechanics, the equivalent quantity would be in agreement with equations 3.29 and 3.31':

$$\Omega = \operatorname{tr} \rho^\dagger \rho = \sum_{nn'} \langle n|\rho|n'\rangle^\dagger \langle n'|\rho|n\rangle$$

$$= \sum_n |\langle n|\rho|n\rangle|^2 + \sum_{n \neq n'} |\langle n|\rho|n'\rangle|^2$$

$$= \sum (\text{diagonal terms})^2 + \sum (\text{off-diagonal terms})^2 \qquad (7.14)$$

with (see equation 3.31″)

$$\operatorname{tr} \rho = \sum_n \langle n|\rho|n\rangle = 1 \qquad (7.15)$$

We may associate the diagonal terms $\langle n|\rho|n\rangle$ with probabilities (i.e., equation 3.31″) and the off-diagonal ones with correlations.

A Lyapounov function of the form 7.13 or 7.14 would indeed incorporate correlations and would go beyond Boltzmann's approach, which deals only with probabilities. We may add that the existence of a Lyapounov function of type 7.14 would be eminently reasonable, because the minimum of Ω, taking into account equation 7.15, would be reached when all diagonal elements of ρ are equal (and their sum equal to one) and all off-diagonal elements vanish. This is the situation described by *equal probabilities* and *random phases*. We would then have a situation quite similar to the microcanonical ensemble, considered in Chapter 2, in which all states have the same probability on the energy surface.

What would happen if we carried out the velocity-inversion experiment using expression 7.14. The result that we could expect is represented in Figure 7.3. (For a detailed discussion, see Prigogine et al., 1973.) Suppose that we start with only diagonal elements in the density matrix (which corresponds to the initial condition of no correlations). We then proceed until time t_0. In this time interval, we have an evolution quite similar to that described by Boltzmann's equation (see Figure 7.2) and Ω

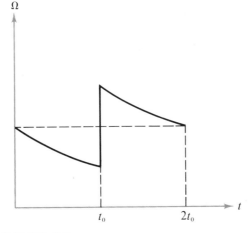
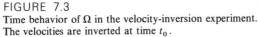

FIGURE 7.3
Time behavior of Ω in the velocity-inversion experiment.
The velocities are inverted at time t_0.

decreases as a result of collisions. At t_0 we obtain a velocity inversion. This corresponds to introducing off-diagonal elements into the density matrix, because such elements correspond to correlations. Therefore, at this point Ω will increase (see expression 7.14); from t_0 to $2t_0$, it will again decrease as the anomalous correlations die out. At time $2t_0$ the system is in the same state as it was at time t_0. In other words, we have restored the initial state at the expense of "entropy production," which is now positive during *all* the time evolution of the system. There is no longer any period of the system corresponding to "antithermodynamic behavior." The increase at time t_0 is not in contradiction with this statement. At this time the system is not closed—the velocity inversion corresponds to a flow in entropy (or of "information"), which leads to an increase. We can contrast this behavior with that of the Boltzmann H-quantity in which the "thermodynamic" evolution from 0 to t_0 is followed by an antithermodynamic one from t_0 to $2t_0$ (see Figure 7.2).

In summary, we may say that we have realized a *cycle* of rejuvenation, but, as in real life, rejuvenation exacts a price. Here, this price is the overall entropy production in the period from 0 to $2t_0$. Can we really construct a function, such as Ω, that takes into account correlations? That is the basic question.

Gibbs Entropy

As just pointed out, we would like to construct a Lyapounov function such as expression 7.13 or 7.14. Let us see if this can be done using the Liouville equation. The calculation is especially simple for classical systems, because we can then obtain (using equation 2.13)

$$\frac{1}{2} \frac{d\Omega}{dt} = -\int \rho(L\rho) \, dp \, dq$$

$$= -\int \rho \left[\frac{\partial H}{\partial p} \frac{\partial \rho}{\partial q} - \frac{\partial H}{\partial q} \frac{\partial \rho}{\partial p} \right] dp \, dq$$

$$= 0 \tag{7.16}$$

which can be easily verified by partial integration. This result is independent of the special functional (expression 7.13). We could also have considered

$$\Omega = \int \rho \log \rho \, dp \, dq \tag{7.16'}$$

or any other "convex" functional of ρ. The attempt to avoid the difficulties in Boltzmann's scheme by considering the complete distribution function ρ instead of the velocity distribution f fails. That is the reason why Gibbs proceeded, as mentioned in Chapter 1, to a "subjectivistic view of irreversibility" as an illusion due to the imperfection of the sensory organs of the observer. (For a recent interpretation of this view see Uhlenbeck in Mehra, 1973.) However, from the point of view adopted in this book, the negative result expressed in equation 7.16 can hardly be a surprise: ensemble theory differs from dynamics in the fact that "ignorance" of initial conditions is incorporated in the distribution function ρ. But this cannot be the *sole* reason why irreversibility, as expressed by a Lyapounov function, can be constructed. Certainly supplementary conditions, such as weak stability, are necessary.

In addition it should not be expected that the Liouville equation (2.12) will lead to a *universal* Lyapounov function, be it expression 7.13 or 7.16'.

Let us consider, instead of the Liouville equation, a system of ordinary linear equations:

$$\frac{d\mathbf{x}}{dt} = A\mathbf{x}$$

Here, too, we may ask whether a Lyapounov function associated with this system of equations exists. This question is discussed in all textbooks dealing with the Lyapounov method for the study of the stability of the solution of differential equations (Minorski 1962). Generally, one considers a quadratic form such as

$$\Omega = \mathbf{x} \cdot \mathbf{Bx} = \sum B_{ij} x_i x_j$$

If Ω is indeed a Lyapounov function, the elements of the term B will in general depend on the coefficients A of the differential equations.

Similarly, we have to expect that, if there is a Lyapounov function associated with dynamics (i.e., with the Liouville equation), it should be a functional of the dynamical processes involved (included in the operator L). A universal form can emerge only at a later stage, through a suitable change of coordinates. Before considering these questions, let us introduce the Poincaré-Misra theorem.

The Poincaré-Misra Theorem

As mentioned early in this chapter, Poincaré reached the conclusion that dynamics and thermodynamics cannot be reconciled. In a sense, this is a direct consequence of his recurrence theorem: "a function of phase space will infinitely often assume its initial value." It therefore cannot behave in a monotonically increasing way as required by the second law. But, by taking a suitable average with distribution functions, this may not be so. Baidyanath Misra (1978) has shown that Poincaré's conclusions are not altered.

The Poincaré-Misra theorem will be presented in a way that directly relates to expression 7.13. Note that we may also write that expression in the form

$$\Omega = \int [e^{-itL}\rho(0)][e^{-itL}\rho(0)] \, dp \, dq$$

$$= \int \rho(0)e^{itL}[e^{-itL}\rho(0)] \, dp \, dq = \int \rho^2(0) \, dp \, dq \qquad (7.17)$$

in which we have used equation 2.12′ and the fact that L is a Hermitian operator (see equation 2.13). We recover the fact that Ω is time independent. We now look for a more general form such as

$$\Omega = \int \rho(t)M\rho(t) \, dp \, dq \geqslant 0 \qquad (7.18)$$

with

$$M \geqslant 0 \qquad (7.19)$$

To make expression 7.18 a Lyapounov function, we suppose that the time derivative D of M is negative (or zero)

$$\frac{dM}{dt} = D \leqslant 0 \qquad (7.20)$$

Using equations 2.5 and 2.13, we may write

$$\frac{dM}{dt} = iLM \qquad (7.21)$$

It is now an easy matter to show that requirement 7.20 cannot be satisfied unless $D = 0$ everywhere, but then Ω is not a Lyapounov function if M is a function of coordinates and momenta. Let us consider the time derivative of Ω:

$$\frac{d\Omega}{dt} = \frac{d}{dt}\left[\int e^{-itL}\rho(0)Me^{-itL}\rho(0) \, dp \, dq\right]$$

$$= -i \int e^{-itL}\rho(0)(LM - ML)e^{-itL}\rho(0) \, dp \, dq$$

$$= \int e^{-itL}\rho(0)De^{-itL}\rho(0) \, dp \, dq \qquad (7.22)$$

We now consider the case corresponding to an equilibrium ensemble (see the section titled Operators in Chapter 2):

$$\rho(0) = \text{microcanonical ensemble} = \text{constant} \qquad (7.23)$$

which we normalize to one. Then by definition

$$\rho(t) = e^{-itL}\rho(0) = \rho(0) \qquad (7.24)$$

and, because equilibrium has been reached, we require that

$$\frac{d\Omega}{dt} = \int \rho(0)D\rho(0) \, dp \, dq = 0 \qquad (7.25)$$

All this is valid when M (and D) are operators or ordinary functions of coordinates and momenta. However, if the latter case is so, we can go one step further and replace $\rho(0)$ in equation 7.25 by its value, which is a constant and which we take equal to one. Then equation 7.25 reduces to

$$\frac{d\Omega}{dt} = \int D \, dp \, dq = 0 \qquad (7.25')$$

But because of expression 7.20 this implies that $D = 0$ everywhere on the microcanonical surface and Ω cannot be a Lyapounov functional. This proof can be extended to general convex functionals. We therefore return to Poincaré's conclusion: the *microscopic* entropy (or Lyapounov functional) cannot be an ordinary function of the phase variables. If it exists at all, it can only be an *operator*. Then equation 7.25 can indeed be satisfied by requiring only that $D\rho(0)$, for $\rho(0) = \text{constant}$, is an eigenfunction of D corresponding to a vanishing eigenvalue. But then the introduction of irreversibility requires a generalization of the conceptual framework of dynamics!

A New Complementarity

What we have shown is not only that the functional of the form 7.13 cannot be used to define a Lyapounov functional—this is a direct consequence of the Liouville equation—but that more general functionals such as those of the form 7.18 are also ruled out if the quantity M corresponding to the "microscopic entropy" is a function of coordinates and momenta.

It should be emphasized that an appeal to special, "improbable" initial conditions would not help. We *assume* the validity of the second law as expressed by a Lyapounov function. We may introduce a weaker statement by giving up the *monotonous* increase of entropy. But then we are lost, for the distinction between reversible and irreversible processes would have to be replaced by some new one, which at present we cannot even formulate in a consistent way. Therefore, it seems that we are back to the difficulties mentioned in Chapter 1. Must we regard irreversibility as an approximation or as a property that we, the observers, introduce into a reversible world? Fortunately, this is not an unavoidable consequence of the Poincaré-Misra theorem. As already explained, since the advent of quantum mechanics, we have become accustomed to introducing into physics a new type of object, *operators* (see Chapters 2 and 3). Therefore it is tempting to consider the Lyapounov functional of the form 7.17, but with M defined as a microscopic "entropy operator" that *does not commute* with the Liouville operator L. The commutator

$$-i(LM - ML) = D \leqslant 0 \tag{7.27}$$

then defines the "microscopic entropy production." But this leads to a new form of complementarity.

The concept of complementarity was introduced in Chapter 3. We have seen that in quantum mechanics position and momenta are represented by noncommuting operators (Heisenberg's uncertainty relations). This may be viewed as an example of Bohr's complementarity principle: there are observables in quantum mechanics whose numerical value cannot be determined simultaneously. Thus, we also have a new form of complementarity—one between the dynamical and the thermodynamic descriptions. The possibility of such a complementarity was explicitly mentioned by Bohr and is confirmed by the approach taken here. Either we consider eigenfunctions of the Liouville operator to determine the dynamical evolution of the system or we consider eigenfunctions of M,

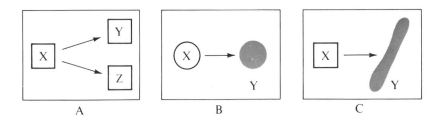

FIGURE 7.4
Three possible transitions of a dynamical system: (A) transition between an initial
phase-space region X at time t_0 and either of two regions Y and Z at a latter time, τ;
(B) a single type of transition from X to Y; (C) distribution of phase fluid initially
concentrated in region X on long filament Y.

but there are no common eigenfunctions of the two noncommuting operators L and M.

What does M regarded as an operator mean? First of all, it means that there are supplementary properties *not* included in the dynamical description. Even if we know the eigenfunctions and the eigenvalues of L, we *cannot* assign a well-defined value to M. Such supplementary properties can come only from some type of randomness in the motion.

We have already seen in Chapter 2 that there is a hierarchy of dynamical systems with stronger and stronger stochastic properties. We have seen that in ergodic systems the motion may be quite smooth (see the section titled Ergodic Systems in Chapter 2), but this is not so when stronger conditions are introduced. Consider a dynamical system that is initially (at time t_0) in region X of phase space. Suppose that at time $t_0 + \tau$ it is found either in region Y or in region Z (see Figure 7.4A). In other words, if we know that at t_0 the system is in region X, we can only calculate the *probability* that it will be in either Y or Z at time $t_0 + \tau$. This does not prove that there is some "basic randomness" associated with the motion. To investigate this point, we decrease the size of region X, in which case one of two things may happen: either for some sufficiently small size of the initial region *all* parts will later be in the "same" region, say Y (see Figure 7.4B), or the situation shown in Figure 7.4A persists *whatever* the size of region X. The second case corresponds precisely to the "weak stability" condition: each region, whatever its size, contains different types of trajectories and the transition to a single trajectory becomes ambiguous.

Our example is somewhat oversimplified: Our requirements are satisfied if each phase element is sufficiently "distorted" with the passage of time. For example, in Figure 7.4C, the phase fluid initially concentrated in region X is distributed after some time on a long filament Y. Again the concept of a trajectory becomes ambiguous if this distortion remains, regardless of the size of region X.

It is in such situations that we may expect a microscopic entropy operator to exist. As will be seen in Chapter 8, this expectation is verified: the operator M can indeed be constructed for systems that present either mixing (or a stronger condition) or a Poincaré catastrophe.

In spite of the basic difference in our arguments, the conception of irreversibility arrived at here is, in its essence, quite similar to that put forward by Boltzmann. Irreversibility is the manifestation *on a macroscopic scale* of "randomness" *on a microscopic scale*.

For examples such as the one just discussed (i.e., that illustrated in Figure 7.4), we may go even further and associate to the system a new type of time—an operator time T closely related to M. Because this T is an operator, it has as eigenvalues the possible ages a system may have (see also Appendix A). A given initial distribution ρ can generally be decomposed into members having different ages and evolving differently.

This is probably the most intriguing conclusion to be drawn in this book: although in physics time was always a mere label associated with trajectories or wave packets, here time emerges with a completely new meaning associated with evolution. We shall return to this idea in Chapters 8 and 9.

It should be emphasized that, although for Boltzmann irreversibility was a consequence of molecular chaos "superimposed" on the equations of dynamics, we pursue a purely dynamical approach. *Both* randomness *and* irreversibility are consequences of the structure of the equations of motion. For example, in classical mechanics we have

Dynamical characteristics $\longrightarrow$ Randomness

(mixing, Poincaré catastrophe) $\searrow$ Irreversibility

                                            (M operator)

Contrary to what Boltzmann attempted to show there is no "deduction" of irreversibility from randomness—they are only cousins!

Chapter 8 will deal first with the consequences of the existence of both the operator M and a Lyapounov function. The construction of the latter will then be discussed briefly, followed by a few examples.

Chapter **8**

THE MICROSCOPIC THEORY OF
IRREVERSIBLE PROCESSES

Irreversibility and the Extension
of the Formalism of Classical
and Quantum Mechanics

We have seen in Chapter 7 that the "minimum assumption" necessary
for introducing irreversibility into classical mechanics is to enlarge the
concept of classical observables: instead of functions of coordinates and
momenta, an *operator* M has been introduced. This means that classical
dynamics no longer consists of the study of orbits; rather it becomes the
study of the time evolution of distribution functions.

The situation is somewhat similar in quantum mechanics. There is no
way of introducing an operator such as M in the framework of the reversible evolution of wave functions as described by the Schrödinger equation (3.17) (see Appendix C).

Therefore we must, as in classical mechanics, turn to ensemble theory
(see Chapter 3) and use the quantum version of the Liouville theorem

Note: This chapter is the most technical one of this book. For the convenience of the
reader, a nontechnical summary is presented in Chapter 9.

(3.36). Moreover, in quantum mechanics we must make a distinction between operators, which act on wave functions, and "superoperators" which act on operators (or matrices). For example, the Liouville operator L acts on the density matrix ρ (see equations 3.35 and 3.36) and is therefore a superoperator.

The entropy operator M in quantum mechanics is also a superoperator because it acts on the density matrix ρ. But it differs in a fundamental way from the Liouville operator L because of the difference between pure states and mixtures introduced in Chapter 3 (see equations 3.30 and 3.32). Described in detail in Appendix C, L is a "factorizable" superoperator, which means that, when acting on ρ corresponding to a pure state (i.e., to a well-defined wave function), it leaves the system in a pure state that is a well-defined wave function. This is in agreement with the Schrödinger equation (3.17), according to which a wave function evolves into another wave function in time. On the other hand, M is *not* factorizable; it does not preserve the difference between pure states and mixtures. In other words, the distinction between pure states and mixtures is lost in systems in which irreversible processes described by a Lyapounov function arise. This does not mean that Schrödinger's equation becomes wrong—nor do Hamilton's equations in classical mechanics—but the distinction between pure states and mixtures (or between wave functions and density matrixes) is no longer observable.

Whenever M can be introduced, we may proceed as in classical mechanics. As usual, the integration over phase space is replaced by the trace operator (see equation 3.32), and expression 7.18 then becomes

$$\Omega = \mathrm{tr}\ \rho^{\dagger} M \rho \geqslant 0 \qquad (8.1)$$

with

$$\frac{d\Omega}{dt} \leqslant 0$$

Again, it is not always possible to find an operator M such that the two preceding inequalities are satisfied. If the Hamiltonian has a discrete spectrum, the motion of the wave function (or of ρ) is periodic. Therefore a necessary condition is the existence of a continuous spectrum.

It is beyond the scope of this book to describe in detail the microscopic theory of irreversible processes. The objective here is simply to assist the reader in grasping the physical meaning of the concepts involved. First, we shall establish the connection between the existence of a Lyapounov function such as expression 8.1 and Boltzmann's approach, and then consider some applications in qualitative terms. We have also seen in Chapter 3 that conventional quantum mechanics has led to unsolved problems being widely discussed today. These problems can be seen from a new perspective once irreversibility is consistently incorporated in the dynamical description.

A New Transformation Theory

Suppose that we were able to construct an operator M for classical or quantum mechanics, such that expression 7.18 or 8.1 represents a Lyapounov function. We would still be far from Boltzmann's ideas, because these Lyapounov functions involve the operator M, which depends on the "dynamics" of the system. In contrast, Boltzmann's H function (equation 7.7) is universal. The remarkable point is that we can use M to generate new, non-Hamiltonian descriptions of dynamics. In the framework of these new descriptions, we may recover the idea of a "universal" H quantity. Indeed, for a closer approach to Boltzmann's ideas, let us represent the entropy operator M as the product of an operator T and its Hermitian conjugate $T^\dagger$. This is always possible because M is positive. (T is the "square root" of M.) We therefore write

$$M = T^\dagger T \qquad (8.2)$$

In keeping with the notation used in earlier publications (see, e.g., Prigogine et al., 1978) we write, instead of T,

$$\Lambda^{-1} \equiv T \qquad (8.3)$$

Inserting definitions 8.2 and 8.3 into expression 8.1, we get, using the definition of Hermiticity (see definitions 3.11 and 3.34''),

$$\Omega = \operatorname{tr} \tilde{\rho}^\dagger \tilde{\rho} \qquad (8.4)$$

with the new transformation density $\tilde{\rho}$ defined by

$$\tilde{\rho} = \Lambda^{-1}\rho \qquad (8.5)$$

This is a very interesting result, because expression 8.4 is of the same type as the one that we were trying to derive from expression 7.14 to describe the velocity-inversion experiment. But we see that this form of the Lyapounov function can exist only in a *new* representation obtained from the preceding one by transformation 8.5. Any explicit reference to the operator M in expression 8.1 has disappeared through the transformation. The definition of a Lyapounov function is not unique. When expression 8.4 is a Lyapounov function, all convex functionals of ρ such as

$$\Omega = \operatorname{tr} \tilde{\rho} \log \tilde{\rho}$$

are also Lyapounov functions (see Appendix A, in which $\tilde{\rho}$ is shown to satisfy a Markov process).

We are dealing with an expression that, like Boltzmann's H-quantity (expression 7.7), depends only on the statistical description of the system. Once we know the state of the system as given by $\tilde{\rho}$, we may evaluate Ω. The particular state $\tilde{\rho}$, which leads to a minimum of Ω, acts as an attractor for the other states. There is therefore a close relation between the existence of the operator M and the transformation theory involving the operator Λ (see definition 8.5).

Let us now reconsider the formal properties of the transformation from expression 8.1 to expression 8.4 (for details, see Prigogine, forthcoming). First we write the equations of motion in the new representation. Taking into account definition 8.5, we obtain

$$i\frac{\partial\tilde{\rho}}{\partial t} = \Phi\tilde{\rho} \qquad (8.6)$$

with

$$\Phi = \Lambda^{-1}L\Lambda \qquad (8.7)$$

The new equation of motion is related to the original one by a similitude (see equation 3.13). But we expect that a transformation that permits the inclusion of "irreversibility" must be more than a mere change of coordinates expressed by a unitary transformation. To clarify this point, we will use the solution of the equations of motion (equation 3.36). We can replace the expressions in 8.1 by the more explicit inequalities

$$\Omega(t) = \text{tr } \rho^\dagger(0)e^{iLt}Me^{-iLt}\rho(0) > 0 \qquad (8.8)$$

$$\frac{d\Omega(t)}{dt} = -\text{tr } \rho^\dagger(0)e^{iLt}i(ML - LM)e^{-iLt}\rho(0) \leqslant 0 \qquad (8.9)$$

We then use expression 8.5 to make the transformation to the new representation and obtain for the entropy production (expression 8.9)

$$\frac{d\Omega}{dt} = -\text{tr } \tilde{\rho}^\dagger(0)e^{i\Phi^\dagger t}i(\Phi - \Phi^\dagger)e^{-i\Phi t}\tilde{\rho}(0) \leqslant 0 \qquad (8.10)$$

This implies that the difference between Φ and its Hermitian adjoint $\Phi^\dagger$ does not vanish:

$$i(\Phi - \Phi^\dagger) \geqslant 0 \qquad (8.11)$$

Therefore we note the important conclusion that the new operator of motion that appears in the transformed Liouville equation (8.6) *can no longer be Hermitian as was the Liouville operator* L. This shows that we must leave the usual class of unitary transformations (expression 3.11) and proceed to an extension of the symmetry of quantum mechanical operators. Fortunately, it is easy to determine the class of transformations that we must consider now. Average values can be calculated in both the old and the new representations. The result should be the same. In other words, we require that

$$\langle A \rangle = \text{tr } A^\dagger \rho = \text{tr } \tilde{A}^\dagger \tilde{\rho} \qquad (8.12)$$

In this sense, the two representations of quantum mechanics should indeed be *equivalent* (if they were not, at least one of them would make incorrect predictions). No experimental information available at present points in this direction.

Moreover, our interest lies in transformations that depend explicitly on the Liouville operator. This is indeed the physical motivation of the theory. We have seen in Chapter 7 that the Boltzmann-type equations have a broken $L - t$ symmetry. We want to realize this new symmetry through our transformation. This can be done only by considering the L-dependent transformation $\Lambda(L)$. The density ρ and observables have the same equations of motion, except that L is replaced by $-L$ (see equations 3.36 and 3.40). We therefore require that, for an observable A,

$$\tilde{A} = \Lambda^{-1}(-L)A \qquad (8.13)$$

Therefore,

$$\operatorname{tr} \tilde{A}^{\dagger}\tilde{\rho} = \operatorname{tr}\{[\Lambda^{-1}(-L)A]^{\dagger}\Lambda^{-1}\rho\}$$
$$= \operatorname{tr}\{A^{\dagger}[\Lambda^{-1}(-L)]^{\dagger}\Lambda^{-1}(L)\rho\} \qquad (8.14)$$

and by identification with the original form of the trace we obtain

$$[\Lambda^{-1}(-L)]^{\dagger} = \Lambda(L)$$
$$\Lambda^{-1}(L) = \Lambda^{\dagger}(-L) \qquad (8.15)$$

which in this development replaces the condition usually imposed on transformations in quantum mechanics, namely that the operators be unitary. If Λ is independent of L, then it is simply a unitary transformation, but this case is of no interest here.

It is not astonishing that we find a *nonunitary* transformation law. Unitary transformations are very much like changes in coordinates, which do not affect the physics of the problem. Whatever the coordinate system, the physics of the system remains unaltered. But now we are dealing with a quite different problem. We want to go from one type of description, the dynamical one, to another, the "thermodynamic" one. This is why we need the drastic type of change in representation expressed by the new transformation law (equation 8.15).

This transformation is called the star-unitary transformation; and a new notation must be introduced:*

$$\Lambda^*(L) = \Lambda^\dagger(-L) \tag{8.16}$$

We shall call this operator the "star-Hermitian" operator associated with Λ ("star" always mean the inversion $L \to -L$ followed by Hermitian conjugation). Then equation 8.15 shows that, for star-unitary transformations, the inverse of the transformation is equal to its star-Hermitian conjugate. As already explained, equation 8.12 can always be satisfied by unitary transformations (they are recovered if we consider Λ independent of L). The remarkable feature is that, in addition, there is a well-defined class of *nonunitary* transformations, which satisfies the equivalence condition and leads to a new form of the equations of motion. Let us now reconsider equation 8.7.

The new dynamical operator Φ is obtained through a similitude from L, but this similitude is in terms of a star-unitary (not a unitary!) operator. Using the facts that L is Hermitian and that equations 8.15 and 8.16 hold, we obtain

$$\Phi^* = \Phi^\dagger(-L) = -\Phi(L) \tag{8.17}$$

or

$$(i\Phi)^* = i\Phi \tag{8.18}$$

The operator of motion is *star-Hermitian*. This is most welcome! Indeed, to be star-Hermitian, an operator may be either Hermitian and even under L-inversion (i.e., it does not change sign when L is replaced by $-L$) or anti-Hermitian and odd (odd means that it changes sign when L is replaced by $-L$). Therefore, a star-Hermitian operator can generally be written as

$$i\Phi = (i\overset{e}{\Phi}) + (i\overset{o}{\Phi}) \tag{8.19}$$

---

* There is an interesting analogy with quantum statistics that may be distinguished by $+1$ or -1 in the distribution functions. Here also the condition of equivalence (equation 8.12) leads to two classes of transformations: $\Lambda^\dagger(L) = \Lambda^{-1}(\pm L)$. The choice of $+$ leads to conventional unitary transformations, whereas $-$ leads to representations displaying irreversible processes.

The superscripts "e" and "o" refer to the even and the odd part of the new time evolution operator Φ. The condition of dissipativity (expression 8.11), which expresses the existence of a Lyapounov function Ω, now becomes

$$i\overset{e}{\Phi} \geqslant 0 \qquad (8.20)$$

It is the even part that gives the "entropy production."

Thus we have obtained a new form of microscopic equation (as was the Liouville equation in classical or quantum mechanics), but our new form explicitly displays a part that can be associated with a Lyapounov function. In other words, the equation

$$i\frac{\partial \tilde{\rho}}{\partial t} = (\overset{o}{\Phi} + \overset{e}{\Phi})\tilde{\rho} \qquad (8.21)$$

contains a *reversible* part and an *irreversible* part. The macroscopic thermodynamic distinction between reversible and irreversible processes has now been incorporated into the microscopic description.

What is so satisfactory here is that the symmetry obtained in equation 8.21 is exactly the *Boltzmann symmetry*. As we have seen in the Boltzmann-type of equation, the collision part is even in L and the flow part, odd.

The physical meaning is also similar. The even term contains all the processes that contribute to the increase of the Lyapounov function and drive the system to equilibrium. This includes scattering, production and decay of particles, damping, and so forth.

The step made through nonunitary transformation is quite crucial. We go from the dynamical description in terms of trajectories or wave packets to a description in terms of processes. It is amazing how the various elements of this approach conspire to achieve a picture that unifies dynamics and thermodynamics. Once we have postulated the existence of the Lyapounov function (expression 8.1), the existence of a representation of dynamics with the characteristic broken "$L - t$ symmetry" follows immediately.

The chain is as follows:

Microscopic entropy operator (M)

> $\rightarrow$ nonunitary transformation Λ
>
> $\rightarrow$ star-Hermitian evolution operator Φ
>
> (with broken symmetry)

In simple cases (such as dilute systems or weakly interacting systems), the new equations of motion have a simple probabilistic interpretation in terms of Markov chains—again in line with Boltzmann's intuition (see Prigogine et al., 1973, and Appendix A). But in our discussion (see Chapter 7) dynamics comes *first*—the physical interpretation including its probabilistic aspect can only be a consequence of the transformation theory.

A posteriori, it is difficult to imagine how the conflict between "being and becoming" could have been resolved in a different way. In the nineteenth century, there was a profusion of controversy between "energeticists" and "atomists," the former claiming that the second law destroys the mechanical conception of the universe, the latter that the second law could be reconciled with dynamics at the price of some "additional assumptions" such as probabilistic arguments. What this means exactly can now be seen more clearly. The "price" is not small because it involves a far-reaching modification of the structure of dynamics.

Construction of the Entropy Operator and the Transformation Theory: The Baker Transformation

So far we have considered only the formal properties of M and its relation to transformation theory. Let us now look briefly into the construction of M and the transformation operator Λ. This in itself is a vast subject and can be dealt with here only in general terms to indicate the methods that must be used (see also Appendixes A and C).

First, we shall consider classical dynamics. Then, as repeatedly mentioned, we must consider two different situations that lead to the type of "weak stability" in which we can expect a Lyapounov function to exist (see Chapter 2).

For ergodic systems, Misra (1978) has shown that mixing is a necessary condition and K-flow a sufficient condition for the existence of the microscopic entropy operator M. As noted in Chapter 2, this classification of dynamical systems is based on the spectral properties of the Liouville operator. Mixing implies that L has no discrete eigenvalues other than zero, and K-flow implies that all eigenvalues of L have the same multiplicity. Note that ergodicity alone is not sufficient; the Liouville operator L must have no discrete eigenvalues other than zero, which corresponds to equilibrium (see Chapter 2), so that there are no periodic motions. Misra has shown that in the case of a K-flow a conjugate Hermitian operator T can be associated with L such that their commutator is constant:

$$-i[L, T] = -i(LT - TL) = 1 \qquad (8.22)$$

in which 1 is the unit operator. A plausibility argument follows (for the proof, see Misra, 1978, and, for an example of the construction, see Appendix A). For a K-flow, we may go to a representation in which the operator L is represented by a number, say λ. We then find an operator T, which in the same representation will be given by the derivative $i(\partial/\partial\lambda)$.

That our approach introduces a new complementarity between dynamics and thermodynamics is especially apparent here because the relation given by equation 8.22 is formally similar to that between momentum and coordinate in quantum theory, as equation 3.2 leads to

$$[q_{op}, p_{op}] = q_{op}p_{op} - p_{op}q_{op} = \hbar i \qquad (8.23)$$

The Liouville operator L corresponds formally to a time derivative (see equation 2.12). Therefore, the conjugate operator T corresponds to a "time" in the sense that the representation

$$L \rightarrow i\frac{\partial}{\partial t}, \qquad T \rightarrow t \qquad (8.24)$$

satisfies the commutation relation (equation 8.22). In other words, we can add to dynamics an operator, T, representing a fluctuating time in accord

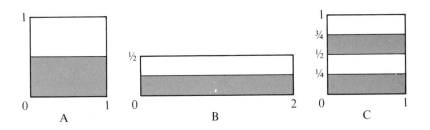

FIGURE 8.1
Baker transformation. First, the unit square (A) is flattened into a
$\frac{1}{2} \times 2$ unit rectangle (B). Then it is reassembled to form a new square
(C) in which the shaded and unshaded areas are split into four
separate regions rather than the two shown in part A.

with the general comments made in Chapter 7. A simple example is
supplied by the *baker transformation*, so-called because it evokes the
image of kneading dough. (This transformation, or mapping, is described
in greater detail in Appendix A.) Consider the unit square shown in
Figure 8.1A. The coordinates x, y are defined by *modulus one*: that is, all
points that do not lie in the unit square are moved into it by adding
integers to or subtracting them from their coordinates. For example,
$(x, y) = (1.4, 2.3)$ is brought into the unit square as $(0.4, 0.3)$.

The transformation is performed at regular time intervals (this is a
discrete transformation):

$$(x, y) \rightarrow (2x, \tfrac{1}{2}y) \qquad \text{mod } 1, \text{ if } 0 \leqslant x < \tfrac{1}{2}$$

$$(x, y) \rightarrow [2x - 1, \tfrac{1}{2}(y + 1)] \qquad \text{mod } 1, \text{ if } \tfrac{1}{2} \leqslant x < 1$$

The mapping has a simple geometrical meaning. If at time t_0 the phase
point is at x, y, then at time $t_0 + \tau$ it is at the point obtained by flattening
the square to a $1/2 \times 2$ rectangle and then cutting and reassembling to
form a new square, as shown in Figure 8.1B, C.

Although this is not a Hamiltonian, dynamical transformation, it can
be used to illustrate many aspects of Hamiltonian flows because it is
measure-preserving. The baker transformation leads precisely to the situation described in Chapter 7 in the section on a new complementarity.
Each finite region is split by the transformation into separate regions.

The operator T in this case has a simple meaning—all of its eigenvalues
are integers from $-\infty$ to $+\infty$. The corresponding eigenfunctions corre-

spond to space distributions that are generated from some standard distribution in a given number of steps. For example, the eigenfunction corresponding to 20, means that 20 applications of the baker transformation are necessary to generate it from the distribution corresponding to the eigenvalue 0. A distribution (more precisely the excess with respect to the uniform equilibrium distribution) may have a well-defined age. It is then by definition an eigenfunction of T. In general, a distribution has no well-defined age but may be expanded in a series of functions having a well-defined age. We may then speak of the average age, of the "fluctuation" of age. The analogy with quantum mechanics is striking. More details can be found in Appendix A.

Once T is known, we can take for M an operator that is a decreasing function of T. We then obtain a Lyapounov function (or an H quantity) that takes its minimum value at microcanonical equilibrium. The meaning of the microcanonical distribution is very simple: whatever the precision of one observation (assuming only that it is *finite*), successive applications of the baker transformation lead to a distribution that is uniform (the inhomogeneity lies below the scale of observation). It is quite remarkable that in such a simple case we may indeed introduce a Lyapounov functional that varies monotonically until the uniform distribution defined in this sense is reached. No thermodynamic limit to a large system is needed.

Moreover, we may, starting from M, introduce a nonunitary transformation Λ to obtain a *universal* Lyapounov function. We write, in agreement with definition 8.5,

$$\tilde{\rho} = \Lambda^{-1}\rho \tag{8.25}$$

with

$$\Lambda^{-1} = M^{1/2}(T) \tag{8.26}$$

We can now see what the L-dependence of $\Lambda(L)$ means. The transformation Λ depends on T, which itself is related to L through the commutation rule (equation 8.22). The L inversion also means the inversion of T:

$$\Lambda(-L) = \Lambda(-T). \tag{8.27}$$

We have seen that Λ is a star-unitary operator satisfying equation 8.15.

Because T and $M(T)$ are Hermitian, this condition reduces to

$$\Lambda^{-1}(L) = \Lambda(-L) \tag{8.28}$$

By inverting L, we obtain the inverse transformation. Such transformations are well known in physics. For example, the Lorentz transformation in special relativity belongs to this class (when we invert the relative velocity between two observers, we obtain the inverse transformation).

That $\tilde{\rho}$ has all the properties of a distribution function (notably it is positive) is verified in Appendix A.* The important point is that to obtain a universal Lyapounov functional, we need to perform a change of variables—a rescaling involving the dynamical properties of the system.

Let us now turn to the second case in which we expect to find weak stability; that is, the Poincaré catastrophe.

Entropy Operator and the Poincaré Catastrophe

The construction of M and Λ is in this case a more arduous task. Interestingly, this is the case that was considered first by the Brussels group (see Prigogine et al., 1973). A survey has been given recently by Alkis Grecos and myself (Prigogine and Grecos 1977). The added difficulty comes from the fact that we need not only the Hamiltonian H (or the Liouville operator L), but the decomposition of H into the "nonperturbed" H_0 and the "perturbation" V (see equation 2.35). This decomposition is most elegantly performed by introducing orthogonal Hermitian projection operators P and Q such that

$$P + Q = 1, P = P^2, Q = Q^2, PQ = QP = 0 \tag{8.29}$$

In terms of these operators,

$$PH = H_0, QH = V \tag{8.30}$$

---

* Moreover, as shown in Appendix A (at least for the class of dynamical systems studied there), we may choose Λ in such a way that $\tilde{\rho}$ satisfies a Markov chain equation. This shows that a statistical scheme may be similar to a dynamical scheme. In other words, the transition from a deterministic to a probabilistic description involves no loss of information.

We can now decompose L, or its resolvent $(L - z)^{-1}$ using these operators. By definition,

$$\frac{1}{L - z} = P\frac{1}{L - z}P + P\frac{1}{L - z}Q + Q\frac{1}{L - z}P + Q\frac{1}{L - z}Q \quad (8.31)$$

Simple manipulations lead to the identity

$$P\frac{1}{L - z}P = \frac{1}{PLP + \Psi(z) - z} \quad (8.32)$$

with

$$\Psi(z) = -PLQ\frac{1}{QLQ - z}QLP \quad (8.33)$$

in which $\Psi(z)$ is the so-called collision operator. It plays a central role in this approach.

The behavior of $\Psi(z)$ for $z \to 0$ is of special interest because it determines the asymptotic behavior of the distribution function [i.e., the limit $\rho(t)$ for $t \to \infty$]. More precisely, it can be shown that traditional kinetic equations such as the Boltzmann equation (or its quantum form, the Pauli equation) can be deduced from the so-called master equation for the N-particle velocity distribution ρ_0, written in the form

$$i\frac{\partial\rho_0}{\partial t} = \Psi(0)\rho_0 \quad (8.34)$$

in which $\Psi(0)$ is the limit of $\Psi(z)$ for $z \to 0$. The existence of kinetic equations is therefore closely related to the nonvanishing of the limit $\Psi(0)$ of the z-dependent collision operator $\Psi(z)$.

The remarkable feature is that $\Psi(0)$ also appears in the theory of the dynamical invariants in connection with Poincaré's theorem. Suppose that the projection operator P projects on the space of the invariants corresponding to the unperturbed motion due to H_0. When we introduce the perturbation V, we hope to "continue" this invariant into a new one, say ϕ, which satisfies condition 2.33 ($L\phi = 0$) and which we now expect to have both a P and a Q part:

$$\phi = P\phi + Q\phi$$

However, using the definition of $\Psi(z)$, it may be shown that this is possible only if the condition

$$\Psi(0)P\phi = 0 \qquad (8.35)$$

is satisfied (see, e.g., Prigogine and Grecos, 1977). If $\Psi(0)$ vanishes, equation 8.35 can evidently always be satisfied and the invariants of H_0 can be extended into invariants of H. On the other hand, when we have what was called in Chapter 2 the "Poincaré catastrophe," the invariants of H_0 cannot be extended into invariants of H (except H itself, or functions of H) and this implies that $\Psi(0)$ is different from zero.

The fact that $\Psi(0)$ appears *both* in kinetic equations of the Boltzmann type (equation 8.34) *and* in the theory of the extension of invariants (equation 8.35) is most important. It shows that Boltzmann's kinetic equations originate *not* in ergodic properties (or stronger properties such as mixing or K-flows) but in the Poincaré catastrophe. This could have been expected. Ergodic theory deals with the Liouville operator only as a *whole*. No decomposition into a part corresponding to free motion (due to the unperturbed Hamiltonian H_0) and to collisions (due to the interaction V) ever appears.

There are however limiting situations such as that for hard spheres in which the potential V is singular (it is either zero or infinite!). Such cases would require that we start with ergodic theory to derive kinetic equations. In spite of much effort, this whole problem is still in a quite preliminary state.

It is also interesting that the condition

$$\Psi(0) \neq 0$$

can be satisfied in very simple systems. Consider, for example, the Hamiltonian

$$H = \omega J + \lambda V \sin \alpha$$

in which ω is an unperturbed frequency and J the action variable for the unperturbed motion ($\lambda = 0$). We calculate $\Psi(0)$ (see Appendix B) and observe that it vanishes for each finite value of λ. However, if we take the limit $\lambda \to \infty$ (or $\omega \to 0$) first, then $\Psi(0)$ does not vanish. This is not surprising: if $\omega = 0$, it is α that becomes a new "action" variable (or constant of motion) and lies in the space orthogonal to the projection operator P. More details are given in Appendix B.

The nonvanishing of $\Psi(0)$ is, however, only a necessary but not a sufficient condition for the construction of operators M or Λ. We need stronger conditions that are related to the behavior of the dispersion equation

$$\Psi(z) - z = 0 \tag{8.36}$$

This equation must admit complex roots. A special method called "subdynamics," has been developed to deal with this problem (see, e.g., Prigogine and Grecos, 1977). A brief example is given in the next section.

In conclusion, it should be emphasized that the construction of the Lyapounov operator M of the nonunitary transformation Λ does not presuppose a single mechanism on the level of the dynamical equation. Various mechanisms may be involved, the important element being that they lead to a complexity on the microscopic level such that the basic concepts involved in the trajectory or the wave function must be superseded by a statistical ensemble.

Microscopic Interpretation of the Second Law of Thermodynamics: Collective Modes

A Lyapounov function that satisfies expressions 8.1 cannot yet be identified with the thermodynamic entropy function. It still corresponds to a purely dynamical concept that may even be applicable to classes of "small" dynamical systems. Also neither M nor Ω is uniquely defined. To identify Ω with macroscopic entropy, supplementary assumptions must be introduced. More precisely, among all irreversible processes only some having a simple macroscopic meaning must be retained. It is indeed a remarkable fact that, among the processes that drive a system to equilibrium, some have remarkable universality and correspond to macro-

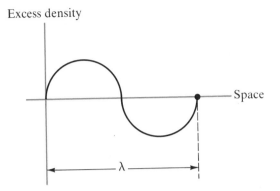

FIGURE 8.2
Excess density as a function of distance.

scopic time scales. These are the so-called hydrodynamic modes, which correspond to the evolution of conserved quantities such as number of particles, momentum, and energy (Forster 1975). This point can be illustrated by means of a system whose density is nonuniform. The excess density is represented in Figure 8.2.

Because a particle cannot disappear (there are no chemical reactions), uniformity will be reached through a slow process of diffusion. The simple Brownian motion model presented in Chapter 1 indicates that the average of the square of the displacement is proportional to time:

$$\langle r^2 \rangle \sim Dt \tag{8.37}$$

We expect that the inhomogeneity will disappear when the distance travelled by the particles is of the order of the wavelength of the perturbation (8.37). As a result, the order of magnitude of the time necessary to destroy the density fluctuation will be

$$\tau \sim \frac{\lambda^2}{D} \tag{8.38}$$

Therefore, this time becomes large when the wavelength increases. This type of process is like those retained in classical hydrodynamics. They are *collective* processes, because they involve a large number of particles (whenever the wavelength is macroscopic). These collective processes include both reversible and irreversible processes such as wave propagation and damping. Therefore equations such as equation 8.21 are quite appropriate because they separate these two parts.

To construct the entropy operator and the transformation function, we must introduce, as in the preceding section, the collision operator $\Psi(z)$, but we must retain only the long time modes in the dispersion equations. This has been done recently by Mary Theodosopulu and Alkis Grecos (1978), who have shown that the Lyapounov function (expression 8.1) then becomes precisely the macroscopic entropy, the Lyapounov function given in equation 4.30 (see Theodosopulu, Grecos, and Prigogine, 1978). Moreover, the moments of the equation of motion (8.21) are the microscopic analogs of the macroscopic hydrodynamic equations.

This is most satisfactory. A bridge between microscopic and macroscopic physics has been achieved. The microscopic Lyapounov function introduced into the dynamical description acquires in this case a direct macroscopic meaning. The only assumptions necessary are short-range forces and small deviations from equilibrium to obtain the linearized equations of hydrodynamics.

Similar results have long been known for dilute gases, starting from the Boltzmann equation. The interesting point is that, in agreement with the expected generality of the second law, nonequilibrium thermodynamics, at least in the linear range, can now be derived from a statistical theory independently of any assumption concerning the density of the system.

Important problems still remain unsolved. We do not yet know if the second law applies to gravitational interactions. Is the second law valid only for a given (or "slowly" varying) gravitational state? Can we include gravitation? We are at the frontier of our knowledge, but it is hoped that, as we begin to understand irreversibility in a more precise way, as a symmetry-breaking mechanism, we will soon be able to make some progress.

We now turn to a basic problem in which the formalism that has been introduced is likely to have interesting applications. As already mentioned, every measurement process includes an element of irreversibility. The measurement must increase the entropy. Thus, it can be seen that the dynamics of the apparatus must be such as to admit the operator M. But we have seen that this requires the concept of weak stability and that in this case the trajectory in the dynamical sense is *no longer an observable* because we cannot extrapolate it from our limited knowledge of phase space.

The complementarity between dynamics and thermodynamics appears here in an especially striking way: either a Lyapounov function exists, in which case the system is not a "pure" dynamical one described by well-defined trajectories but only by statistical distribution functions; or no

Lyapounov function exists, and the system is described by trajectories.

Nevertheless, as indicated in Appendix C, the main conclusion remains true: quantum systems for which the microscopic entropy operator M may be defined are such that the distinction between pure states and mixtures is lost.

Particles and Dissipation: A Non-Hamiltonian Microworld

As mentioned earlier, the interest of equation 8.21 is its direct connection with the second law through inequality 8.20. This connection has relevance to a basic question that remains unanswered in spite of all the work that has been done on it. How is the concept of an elementary particle related to that of interaction?

Taking the example of interacting electrons and photons mentioned in Chapter 3, we generally start by using a Hamiltonian involving the "bare" particles (electrons and photons) and an interaction. These "bare" particles cannot be the "physical" ones. Because of the electro-magnetic interaction between electrons and photons, an electron is always surrounded by a cloud of photons. The bare electron (without photons) is only a formal concept. We then perform a "renormalization process" in which part of the interaction is used to change the physical characteristics of the particle such as its mass or its charge. But at what point do we stop this process? Even after the system has been renormalized, we are faced with the "Hamiltonian dilemma": either no well-defined particles (because the energy is partly "between" the electrons and the photons) or noninteracting particles (in the representation in which the total Hamiltonian is diagonal).

Is there a way out? The important point is that we now have a third description in terms of processes (see Figures 2.5 and 8.3). The electrons and photons are involved in physical processes such as scattering, photon emission, and absorption. These processes drive the total system (electrons plus photons) to equilibrium. Moreover, these processes are "real"; they are part of the evolution of the physical universe. They shall certainly *not* be transformed away by any change of representation. Therefore, whatever the description may be, it should be obtained through a star-unitary transformation leading to dissipativity condition 8.20.

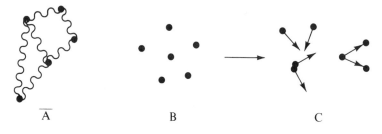

FIGURE 8.3
Three descriptions of a system: (A and B) the two Hamiltonian views; (C) description in terms of processes.

But this cannot be enough; there are families of star-unitary transformations—all of which satisfy condition 8.20. Which one to choose is a problem quite similar to that of the Born–Heisenberg–Jordan quantization rules mentioned in Chapter 3. The latter can be solved by considering all *unitary* transformations and choosing the one that leads to a diagonal form of the Hamiltonian operator. Here, too, we need a quantum rule, but a new one for choosing between the star-unitary transformations. How such a rule may be formulated will be described next; as could be expected, it will be in terms of superoperators.

Remember that the Liouville operator corresponds to a commutator (see equation 3.35),

$$L\rho = H\rho - \rho H \tag{8.39}$$

But we may also introduce an "anticommutator,"

$$\mathscr{H}\rho = \tfrac{1}{2}[H\rho + \rho H] \tag{8.40}$$

The two quantities L and $\mathscr{H}$ are *superoperators* (remember that usual operators act on wave functions, whereas L and $\mathscr{H}$ act on operators). The average value of the energy can be written in terms of the new quantity $\mathscr{H}$ as (see equation 3.38)

$$\langle H \rangle = \operatorname{tr} H\rho = \tfrac{1}{2}\operatorname{tr}(H\rho + \rho H) = \operatorname{tr} \mathscr{H}\rho \tag{8.41}$$

We now apply our transformation Λ to both L and $\mathscr{H}$. In addition to obtaining equation 8.7, we obtain

$$\tilde{\mathscr{H}} = \Lambda^{-1}\mathscr{H}\Lambda \tag{8.42}$$

We now look for a Λ such that condition 8.20 is satisfied and that in addition equation 8.41 can be written as

$$\langle H \rangle = \sum_i \tilde{E}_i \tilde{\rho}_{ii} \qquad (8.43)$$

When this is so, the $\tilde{E}_i$ may be regarded as the energy levels associated with the system. We then have a most satisfactory description of the system: it evolves in accord with the second law (inequality 8.20) and yet the particles can be characterized by well-defined energies.

The method that we have used can be summarized as follows. In conventional quantum mechanics, both the energy levels (see equation 3.16) and the time evolution (equation 3.17) are determined by the same quantity, the Hamiltonian operator H_{op}. This is a kind of remarkable "degeneracy" characteristic of quantum mechanics. However, after the Λ-transition, the superoperator formalism allows us to obtain two different operators: Φ for the time evolution (see expression 8.2), and $\mathscr{H}$ for the determination of the energy levels. In this way, this degeneracy is lifted for systems for which a star-unitary transformation Λ leading to a Lyapounov representation can be defined.

The method is quite new (Prigogine and George 1978; George et al. 1978). It has been applied successfully to a very simple model (the "Friedrichs' model"), but its generality has yet to be investigated. The reasons for mentioning it here are that it avoids the technical difficulties mentioned in Chapter 3, and we obtain strictly exponential decay (the lifetime is a matrix element of $\overset{e}{\Phi}$). But in addition it is the whole concept of "elementary particles" that is at stake!

The classical order was: particles first, the second law later—being before becoming! It is possible that this is no longer so when we come to the level of elementary particles and that here we must *first* introduce the second law before being able to define the entities. Does this mean becoming before being? Certainly this would be a radical departure from the classical way of thought. But, after all, an elementary particle, contrary to its name, is not an object that is "given"; we must construct it, and in this construction it is not unlikely that *becoming*, the participation of the particles in the evolution of the physical world, may play an essential role.

THE BELOUSOV-ZHABOTINSKII REACTION: CHEMICAL SCROLL WAVES

Spiral chemical waves develop when the Belousov-Zhabotinskii reagent is allowed to stand in a shallow dish. The waves can appear spontaneously or be initiated by touching the surface with a hot filament, as in the photographs above. The small circles are bubbles of carbon dioxide evolved by the reaction (see the section on coherent structures in chemistry and biology in chapter 5). After the initial photograph was taken, subsequent ones were taken at 0.5, 1.0, 1.5, 3.5, 4.5, 5.5, 6.5, and 8.0 seconds. Photographs by Fritz Goro.

Chapter 9

THE LAWS OF CHANGE

Einstein's Dilemma

I am writing this chapter in 1979, Einstein's centennial year. Nobody has made greater contributions to the statistical theory of matter and more specifically to the theory of fluctuations than Einstein. Through the inversion of Boltzmann's formula (1.10), Einstein derived the probability of a macroscopic state in terms of the entropy associated with it. This step has proved to be decisive for the whole macroscopic theory of fluctuations (of special interest near critical points). Einstein's relation is a basic element in the proof of the Onsager reciprocity relations (equation 4.20).

Einstein's description of brownian motion, as summarized in Chapter 1, was one of the first examples of "random processes." Its interest is far from being exhausted even today. The modeling of chemical reactions by Markov chains, described in Chapter 6, is an extension of the same line of thought.

Finally, it was Einstein who first recognized the general meaning of Planck's constant h as leading to the wave-particle duality. Einstein was concerned with electromagnetic radiation. But about twenty years later, de Broglie extended Einstein's relations to matter. The work of Heisenberg, Schrödinger, and others put these ideas into a mathematical framework. But, if matter is both wave and particles, the idea of trajectory of classical determinism is lost. As a result, only statistical predictions can be made by quantum theory (see Chapter 3 and Appendix D). To the end of his life Einstein remained opposed to the idea that such statistical considerations correspond to objective features of nature. In his well-known letter to Max Born (see Einstein, 1969), he wrote:

> You believe in the God who plays dice, and I in complete law and order in a world which objectively exists, and which I, in a wildly speculative way, am trying to capture. I firmly *believe*, but I hope that someone will discover a more realistic way, or rather a more tangible basis than it has been my lot to do. Even the great initial success of the quantum theory does not make me believe in the fundamental dice game, although I am well aware that your younger colleagues interpret this as a consequence of senility.

Why did Einstein take such a strong view concerning time and randomness? Why did he prefer intellectual isolation to any compromise in these matters?

Among the most moving documentation of Einstein's life is the collection of letters that he exchanged with his old friend Michele Besso (Einstein 1972). Einstein was usually very reticent about himself, but Besso was a very special case. They knew each other at an early age in Zurich when Einstein was seventeen and Besso twenty-three. Besso took care of Einstein's first wife and their children in Zurich when Einstein was working in Berlin. Although the affection between Besso and Einstein remained deep, their interests diverged with the years. Besso became more and more involved in literature and philosophy—in the very meaning of human existence. He knew that, to obtain a response from Einstein, he had to include problems of a scientific nature, but his interest was more and more elsewhere. Their friendship lasted their whole lives, Besso having died only a few months earlier than Einstein in 1955. It is mainly the last part of the correspondence between 1940 and 1955 that is of interest to us here.

There Besso returned again and again to the problem of time. What is irreversibility? How does it relate to the basic laws of physics? And patiently Einstein answered again and again, irreversibility is an illusion, a subjective impression, coming from exceptional initial conditions. Besso remained dissatisfied. His last scientific paper was a contribution to the *Archives des Sciences* published in Geneva. At the age of eighty, he presented an attempt to reconcile general relativity and irreversibility of time. Einstein was not happy with this attempt: "You are on a gliding ground," he wrote. "There is no irreversibility in the basic laws of physics. You have to accept the idea that subjective time with its emphasis on the now has no objective meaning." When Besso passed away, Einstein wrote a moving letter to his widow and son: "Michele has preceded me a little in leaving this strange world. This is not important. For us who are convinced physicists, the distinction between past, present, and future is only an illusion, however persistent."

Einstein believed in the god of Spinoza, a god identified with nature, a god of supreme rationality. In this conception there is no place for free creation, for contingency, for human freedom. Any contingency, any randomness that seems to exist is only apparent. If we think that our actions are free, this is only because we are ignorant of their true causes.

Where do we stand today? I believe that the main progress that has been accomplished is that we begin to see that probability is not necessarily associated with ignorance, that the distance between deterministic and probabilistic descriptions is less great than most contemporaries of Einstein and Einstein himself were believing. Poincaré (Poincaré 1914) had already pointed out that, when we throw dice and use probabilities to predict the outcome, we do not mean that the concept of trajectories doesn't apply. Rather, the type of system is such that in each interval of initial conditions, as small as we want, the same number of trajectories go to each side of the dice. This is a simple version of the problem of dynamic instability that has been discussed repeatedly (see Chapters 2, 3, 7, and 8). Before returning to it once again, let us take an overview of the laws of change that have been described.

Time and Change

In Chapter 1, I presented the methods developed through the decades to describe change. Basically, we may distinguish three classes: macroscopic methods dealing with the evolution of averages such as Fourier's law, chemical kinetics, and so forth; stochastic methods such as Markov chains; and classical or quantum mechanics.

Some very unexpected features have emerged in recent years. First, the unexpected wealth of the macroscopic description, especially for nonlinear, far-from-equilibrium situations. This is well illustrated by the reaction-diffusion equations considered in Chapter 5. Even simple examples may lead to successive bifurcations and various time-space structures. This drastically limits the unifying power of macroscopic description and shows that it cannot by itself provide us with a consistent description of time evolution. Indeed, all the various branches represented in Figure 5.2 satisfy the appropriate boundary conditions (in contrast with classical problems in potential theory in which for given boundary conditions there exists a unique solution). In addition, macroscopic equations do not yield information about what happens at the bifurcation points. What will the fraction of systems be, following a given history of bifurcations?

We must therefore turn to stochastic theory such as Markov chains. But here also new features appear. Of special interest is the close relation between fluctuations and bifurcations (see Chapter 6) which leads to deep alterations in the classical results of probability theory. The law of large numbers is no longer valid near bifurcations and the unicity of the solution of linear master equations for the probability distribution is lost (see the section on nonequilibrium phase transitions in Chapter 6).

Yet the relation between stochastic and macroscopic methods is clear. It is precisely when the average quantities do not satisfy closed equations, which happens near bifurcation points, that we must use the full apparatus of statistical theory. The relation between macroscopic or stochastic methods and dynamical ones, however, remains a challenging problem. This question has been considered in the past from many points of view. For example, in his beautiful book *The Nature of the Physical World*

(1958, p. 75), Arthur Eddington introduced a distinction between "primary laws," controlling the behavior of single particles, and secondary laws, such as the principle of the increase of entropy, which would be applicable only to collections of atoms or molecules.

Eddington fully recognized the importance of entropy. He wrote (p. 103): "From the point of view of philosophy of science the conception associated with entropy must, I think, be ranked as the great contribution of the nineteenth century to scientific thought. It marked a reaction from the view that everything to which science need pay attention is discovered by a microscopic dissection of objects."

How can "primary" laws coexist with "secondary" ones? "One would not be surprised," Eddington wrote (p. 98), "if in the reconstruction of the scheme of physics, which the quantum theory is now pressing on us, secondary laws become the basis and primary laws are discarded."

Certainly quantum theory plays a role because it forces us to give up the idea of classical trajectories. But from the point of view of the relation with the second law, the concept of instability, which has been repeatedly discussed, seems to be of fundamental importance. The structure of the equations of motion with "randomness" on the microscopic level then emerges as irreversibility on the macroscopic level. In this sense, the meaning of irreversibility was already anticipated by Poincaré (1921), who wrote:

> In conclusion, using ordinary language, the law of conservation of energy (or the principle of Clausius) can have only one significance, which is that there is a property common to all the possibilities; but on the deterministic hypothesis there is only a single possibility, and the law has no longer any meaning. On the indeterministic hypothesis, on the other hand, it would have a meaning, even if it were taken in an absolute sense; it would appear as a limitation imposed upon freedom. But these words remind me that I am digressing and am on the point of leaving the domains of mathematics and physics.

Poincaré's confidence in a basic deterministic description was too firmly established to consider seriously a statistical description of nature. The situation is quite different for us. Many years after the foregoing passage was written, our confidence in the deterministic description of nature has been shaken both at the microscopic level and at the macroscopic one. We no longer recoil in horror from such bold conclusions!

Moreover, we see that in a sense our point of view reconciles the conclusions of Boltzmann and Poincaré. Boltzmann, the daring revolutionary physicist, whose thought was based on an extraordinary physical intuition, guessed the type of equation that could describe evolution of matter on the microscopic level and still display irreversible processes. Poincaré, with his deep mathematical insight, could not be satisfied with only intuitive arguments, but he clearly saw the only direction in which a solution could be found. It is my belief that the methods summarized in this book (see Chapters 7 and 8 and the Appendixes) constitute the link between Boltzmann's great intuitive work and Poincaré's requirement of mathematization.

This mathematization leads us to a new concept of time and irreversibility to which we now turn.

Time and Entropy as Operators

Much of Chapter 7 dealt with some of the most significant attempts made in the past to define entropy on the microscopic level, with emphasis on Boltzmann's fundamental contribution to this subject, culminating in his discovery of the H-function (equation 7.7). However, independently of other remarks, Boltzmann's H-theorem could, in line with observations presented by Poincaré, not be considered to be "derived" from dynamics. Boltzmann's kinetic equation, on which the derivation of the H-theorem is based, does *not* share the symmetry of classical dynamics (see the section titled Boltzmann's Kinetic Theory in Chapter 7 and that titled A New Transformation Theory in Chapter 8). In spite of its historical importance, it can at most be considered a phenomenological model.

Ensemble theory does not lead us further even when extended by associating entropy with a microscopic phase function (in classical mechanics) or a Hermitian operator (in quantum mechanics). These negative conclusions are described in the sections titled Gibbs Entropy and the Poincaré-Misra Theorem in Chapter 7.

This left us with very few possibilities, short of accepting the view that irreversibility results from mistakes or from supplementary approximations added to classical or quantum mechanics.

However another, radically different approach has now emerged: the idea of associating macroscopic entropy (or Lyapounov function) with a microscopic entropy operator called M.*

This is a momentous step: we were accustomed to considering, in classical mechanics, observables to be functions of correlations and momenta. Yet the introduction of the Liouville operator L in both classical and quantum ensemble theory (see Chapters 2 and 3) has prepared us for this new step, which is of a quite different nature. Indeed, ensemble theory was considered to be an "approximation," whereas the "basic" theory was in terms of trajectories or wave functions. With the introduction of operator M, the situation becomes quite different. It is the description in terms of bundles of trajectories, or distribution functions, that becomes basic; no further reduction to individual trajectories or wave functions can be performed.

The physical meaning of entropy and time as operators is discussed in Chapters 7 and 8, as well as in Appendixes A and C. (See especially the introduction and the section titled An Extended Complementarity Principle in Chapter 7 and the sections titled Irreversibility and the Formalism of Classical and Quantum Mechanics, The Construction of the Entropy Operator and the Transformation Theory, and Particles and Dissipation—A Non-Hamiltonian Microworld in Chapter 8. Because operators were first introduced in physics through quantum mechanics, there remains in the minds of most scientists a close relation between quantization involving Planck's constant h and the appearance of operators. The association of operators with physical quantities has, however, a broader meaning quite independent of quantization. It means basically that for some reason the classical description in terms of trajectories has to be given up either because of instability and randomness on the microscopic level (see Appendix A) or because of quantum "correlations" (see Appendix D).

---

* From a historical point of view, it is interesting that the nonunitary transformation Λ, which leads from the usual Liouville equation to the kinetic equations (see the section titled A New Transformation Theory in Chapter 8), was found first. It has only recently been realized that this means that a supplementary operator M exists in the original (Hamiltonian) representation and therefore in this sense the usual dynamical description was not complete.

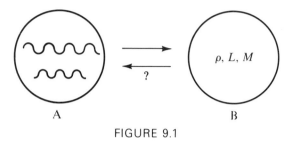

FIGURE 9.1

For classical mechanics we may present the situation in the following way. The usual description (Figure 9.1A) is in terms of trajectories or orbits generated by Hamilton's equations (2.4). The other description (Fig. 9.1B) is in terms of distribution functions (2.8), their motion being determined by the Liouville operator.

These two descriptions can be different only if we cannot at each moment go from one description to the other. The physical reasons for this are discussed in the section on weak stability in Chapter 2. Experiments performed with an arbitrary but finite accuracy lead us only to the identification of some *finite* region of phase space where the system may be located. The question is then whether we can perform, at least in principle, a transition-limiting process, as indicated schematically in Figure 9.2, from this region to a point P, to a δ-function corresponding to a well-defined orbit.

This transition-limiting process is related to the question of weak stability discussed in Chapter 2. It becomes impossible to perform when we have a variety of trajectories in each region of phase space—*however small*. Then the microscopic description becomes so "complex" that we cannot go beyond it in terms of distribution functions. At present, we know of two types of dynamical systems for which this is so—systems with sufficiently strong mixing properties and systems presenting the Poincaré catastrophe (see Chapters 2 and 7 and Appendixes A and B). In fact, almost "all" dynamical systems with the exception of a few "school" examples belong to these categories. We shall return to this question in the next section.

One might think that such "natural limits" of classical or quantum physics would lead to a decrease of their predictive power. In my opinion, the reverse is true. We can now make statements about the evolution of distribution functions that go beyond what can be said about individual trajectories. New concepts appear.

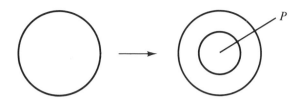

FIGURE 9.2

Among these new concepts some of the most interesting are the microscopic entropy operator M and the time operator T. Here we are dealing with a second time, an internal time quite different from the time that in classical or quantum mechanics simply labels trajectories or wave functions. We have seen that this operator time satisfies a new uncertainty relation with the Liouville operator L (see equation 8.22 and Appendixes A and C). We may define averages $\langle T \rangle$, $\langle T^2 \rangle$ through the bilinear forms

$$\langle T \rangle = \mathrm{tr}\, \rho^\dagger T \rho, \quad \langle T^2 \rangle = \mathrm{tr}\, \rho^\dagger T^2 \rho \tag{9.1}$$

Interestingly enough, the "ordinary" time—the label of dynamics—then becomes an average over the new operator time. This is in fact a consequence of the uncertainty relation (8.22), which implies that

$$\frac{d}{dt} \langle T \rangle = \frac{d}{dt}\, \mathrm{tr}[(e^{-iLt}\rho)^\dagger T e^{-iLt}\rho]$$

$$= i\, \mathrm{tr}[\rho^\dagger e^{iLt}(LT - TL)e^{-iLt}\rho]$$

$$= \mathrm{tr}\, \rho^\dagger \rho = \text{constant} \tag{9.2}$$

With an appropriate normalization we may take this constant equal to one. We see therefore that

$$dt = d\langle T \rangle \tag{9.3}$$

In other words, macroscopic time is simply the average over the new operator time. In this perspective, the usual time concept is recovered only when T becomes a trivial operator such that (in classical mechanics)

$$T\rho(x, v, t) = t\rho(x, v, t) \tag{9.4}$$

Then "age" is independent of the form of the distribution in phase space.

On the contrary, the new concept implies that age depends on the distribution itself and is therefore no longer an external parameter, a simple label as in the conventional formulation (see Appendix A).

We see how deeply the new approach modifies our traditional view of time, which emerges now as a kind of average over "individual times" of the ensemble.

Levels of Description

For a long time, the absolute predictability of classical mechanics, or the physics of being, was considered to be an essential element of the scientific picture of the physical world. It is quite remarkable that over the three centuries of modern science (it seems indeed legitimate to consider 1685, the year Newton presented his *Principia* to the Royal Society, as the birth date of modern science) the scientific picture has shifted toward a new, more subtle conception in which both deterministic features and stochastic features play an essential role.

Let us consider only the statistical formulation of the second law of thermodynamics by Boltzmann, in which the concept of probability played an essential role for the first time. We then have quantum mechanics, which preserves determinism but in the framework of a theory that deals with wave functions having a probabilistic content. In this way, probabilities appeared for the first time in the basic, microscopic description.

This evolution is still continuing. We find essential stochastic elements not only in the theory of bifurcations on the macroscopic level (see Chapter 5), but also in the microscopic description as provided even by classical mechanics (see Chapters 7 and 8). As we have seen, these new elements lead finally to new concepts for time and entropy, the consequences of which must yet be explored.

It is remarkable that classical dynamics, statistical mechanics, and quantum theory can be discussed starting from the ensemble point of view introduced by Einstein and Gibbs. When the transition from an ensemble to a single trajectory can no longer be performed, we obtain different theoretical structures. Classical dynamics has been discussed herein from this point of view, especially the transition to statistical mechanics as a result of weak stability. It has also been mentioned that the existence of the universal constant h introduces correlations in phase space and prevents the transition from ensembles to single trajectories (further details are given in Appendixes C and D). The results are depicted in the following scheme.

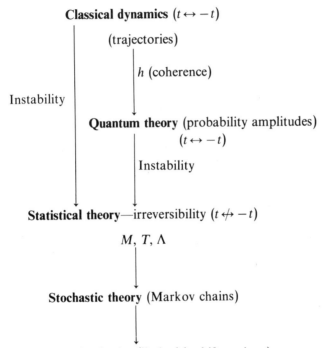

We begin to be able to coordinate the various levels of description repeatedly discussed in this book. However, a few words of caution are necessary. For example, we may transform the deterministic description (in terms of a Liouville equation) into a Markov chain (see Appendix A) for a class of strongly unstable systems to which the baker transformation belongs. Is this so in more general situations involving a weaker form of instability such as the mixing property considered in Chapter 2? Another example is the quantum mechanics theory (see Appendix C). Quantum mechanical instability theory is still in its infancy.

Supplementary classifications and new points of view are likely to emerge in the future. Yet our present scheme is not empty and brings some unifying features into the structure of theoretical physics.

A few comments seem appropriate here on the dynamical complexity associated with instability. In classical dynamics, some simple situations that are time reversible ($t \leftrightarrow -t$) can at least be conceived of. Whenever chemical processes (and a fortiori whenever biological processes) are considered this becomes impossible because chemical reactions are always associated—nearly by definition—with irreversible processes. Moreover, measurements—which extend our sensory perceptions— necessarily involve some element of irreversibility. Therefore, the two formulations of the laws of nature (one for which $t \leftrightarrow -t$ and the other for which $t \nleftrightarrow -t$) are equally fundamental. We need both. It is true that we may consider the world of trajectories (or of wave functions) to be the fundamental one. From this perspective, new formulations are obtained when supplementary assumptions are introduced. But we can also consider irreversibility to be a basic element of our description of the physical world. From this perspective, the world of trajectories and wave functions corresponds on the contrary to idealizations of great importance, but they lack essential elements and cannot be studied in isolation.

We have arrived at a kind of self-consistent picture which will be described in a little more detail.

Past and Future

Once we can add a Lyapounov function to the dynamics, future and past can be distinguished, exactly as in macroscopic thermodynamics in which the future is associated with a larger entropy. But again some caution is necessary. We may construct a Lyapounov function that increases mono- tonously with the "flow" of time or another one that decreases. In more technical terms, the transition from the situation represented in Figure 9.1A, which corresponds to a dynamical *group*, to the one represented in Figure 9.1B, which corresponds to a *semigroup*, can be performed in two ways: in one description equilibrium is reached in the "future," and in the other in the "past." In other words, the time symmetry of dynamics can be broken in two ways; however, how to distinguish between them is a difficult question. As emphasized in the preceding section, life even in its simplest form *presupposes* a distinction between past and future. Monocellular organisms such as amoebas move from media poor in nutrients to media rich in nutrients. Even such organisms *anticipate* the future through signals received from their environment.

When we study time-reversible laws of dynamics, we make a distinction between past and future—between, say, predicting the position of the moon or calculating what its position was in the past. The distinction between past and future is a kind of *primitive concept* that in a sense precedes scientific activity. We may, however, include this primitive concept in a self-consistent scheme, as shown in the following diagram:

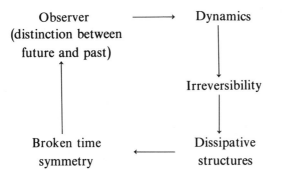

We start with the observer, a living organism who makes the distinction between the future and the past, and we end with dissipative structures, which contain, as we have seen, a "historical dimension." Therefore, we can now recognize ourselves as a kind of evolved form of dissipative structure and justify in an "objective" way the distinction between the future and the past that was introduced at the start.

Again there is in this view no level of description that we can consider to be the fundamental one. The description of coherent structures is not less "fundamental" than is the behavior of the simple dynamical systems.

Note that the transition from one level to the other involves "symmetry-breaking"; the existence of irreversible processes on the microscopic level as described through kinetic equations violates the symmetry of canonical equations (see Chapter 8), and dissipative structures may in turn break the symmetries of space-time.

The very possibility of such a self-consistent scheme implies the existence of nonequilibrium processes and therefore a picture of a physical universe that for some cosmological reasons provides the necessary type of environment. Although the *distinction* between reversible and irreversible processes is a problem of dynamics and does not involve cosmological arguments, the possibility of life, the activity of the observer, cannot be dissociated from the cosmological environment in which we happen to be. However, the questions, What is irreversibility on the cosmic scale? Can we introduce an entropy operator in the framework of a dynamical description in which gravitation plays an essential role? are formidable ones. I prefer to confess my ignorance.

An Open World

The basis of the vision of classical physics was the conviction that the future is determined by the present, and therefore a careful study of the present permits the unveiling of the future. At no time, however, was this more than a theoretical possibility. Yet in some sense this unlimited predictability was an essential element of the scientific picture of the physical world. We may perhaps even call it the founding myth of classical science.

The situation is greatly changed today. It is remarkable that this change results basically from our better understanding of the limitations of measurement processes because of the necessity to take into account the role of the observer. This is a recurrent theme in most of the basic ideas that originated with the development of physics in the twentieth century.

It was already present in Einstein's analysis of space-time (1905) in which the limitation of the speed of propagation of signals to velocities smaller than the velocity of light in a vacuum plays such an essential role. It is certainly not *logically* inconsistent to suppose that signals may be transmitted with infinite speed, but this Galilean space-time concept seems to conflict with a whole host of experimental information that has been gathered through the years. The incorporation of the limitation of our way of acting on nature has been an essential element of progress.

The role of the observer in quantum mechanics has been a recurrent theme in the scientific literature in the past fifty years. Whatever the future developments are, this role is essential. The naive realism of classical physics, which assumed that properties of matter were "there" independently of the experimental device, had to be revised.

Again, the developments described in this book point in a similar direction. Theoretical reversibility arises from the use of idealizations in classical or quantum mechanics that go beyond the possibilities of measurement performed with any finite precision. The irreversibility that we observe is a feature of theories that take proper account of the nature and limitation of observation.

At the origin of thermodynamics we find "negative" statements expressing the impossibility of certain transformations. In many textbooks, the second law of thermodynamics is expressed as the postulate that it is impossible to transform heat into work using a single thermostat. This negative statement belongs to the macroscopic world—in a sense we have followed its meaning to the microscopic level when it becomes, as we have seen, a statement about the observability of the basic conceptual entities of classical or quantum mechanics. As in relativity, a negative statement is not the end of the story: it leads in turn to new theoretical structures.

Have we lost essential elements of classical science in this recent evolution? The increased limitation of deterministic laws means that we go from a universe that is closed, in which all is given, to a new one that is open to fluctuations, to innovations.

For most of the founders of classical science—even for Einstein—science was an attempt to go beyond the world of appearances, to reach a timeless world of supreme rationality—the world of Spinoza. But perhaps there is a more subtle form of reality that involves both laws and games, time and eternity. Our century is a century of explorations: new forms of art, of music, of literature, and new forms of science. Now, nearly at the end of this century, we still cannot predict where this new chapter of human history will lead, but what is certain at this point is that it has generated a new dialogue between nature and man.

APPENDIXES

TIME AND ENTROPY OPERATORS FOR THE BAKER TRANSFORMATION

The following discussion is an attempt to explain how the time operator T (see equation 8.22) and a microscopic entropy operator M may be associated with the baker transformation introduced in Chapter 8.

The results given here summarize a recent paper by Misra, Prigogine, and Courbage[1] in which all proofs, as well as various generalizations of the results to other systems, can be found. Other aspects related to the baker transformation that are not treated here can be found in important papers by Lebowitz, Ornstein, and others.[2, 3, 4, 5, 6]

Phase space Ω will be the unit square in the plane. As shown in Figure 8.11, the baker transformation, B, sends point $\omega = (p, q)$ of Ω into $B\omega$, with

$$B\omega = (2p, q/2) \quad \text{if } 0 \leqslant p < \tfrac{1}{2}$$

$$B\omega = (2p - 1, q/2 + \tfrac{1}{2}) \quad \text{if } \tfrac{1}{2} \leqslant p < 1 \tag{A.1}$$

Transformation B describes a discrete process that takes place at regular time intervals and tends to progressively fragment an arbitrary given

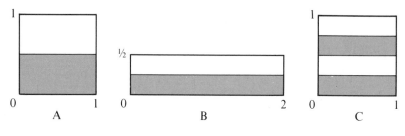

FIGURE A.1
Application of the baker transformation to the half square.

surface element. As an example, let us apply transformation B to the half square $0 \leqslant q \leqslant \frac{1}{2}$. The result is shown in Figure A.1.

When the baker transformation is repeated many times, the initial half square is broken into smaller and smaller rectangles as shown in Figure A.2.

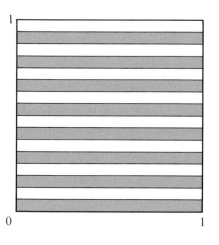

FIGURE A.2
Effect of successive baker transformations on the half square.

After some time, the fragmentation becomes so fine that whatever the precision of observation (supposed only to be finite) the distribution will appear uniform. At this stage the system has reached its equilibrium (microcanonical) distribution.

The baker transformation admits a remarkable representation as a "Bernouilli shift." To understand this relation, we write the coordinates p

and q in a binary expansion:

$$p = 0.u_1 u_2 \cdots$$

$$q = 0.u_0 u_{-1} \cdots$$

This notation means that

$$p = \frac{u_1}{2} + \frac{u_2}{2^2} + \cdots$$

and similarily for q. Here the u_i's take the values 0 or 1.

A point ω in Ω is therefore represented by the double sequence $\{u_i\}$ with $i = 0, \pm 1, \pm 2 \cdots$. Using specific examples, one can easily verify that to $B\omega$ there corresponds the sequence $\{u_i'\}$, in which $u_i' = u_{i+1}$. We see clearly that the baker transformation induces a shift in the sequence. It is mainly for this reason that one speaks of a "Bernouilli shift."[6]

Consider a simple orthonormal basis for all square integrable functions on the "phase space." Let X be the function defined on $(0, 1)$ by

$$X(1) = 1$$
$$X(0) = -1 \tag{A.2}$$

For each integer n, a function $X_n(\omega)$ is defined on Ω by

$$X_n(\omega) = X(u_n) \tag{A.3}$$

The value of $X_n(\omega)$ in each point of Ω therefore depends solely on the nth digit in the binary expansion of the coordinates p, q.

In addition, we define for each finite set of integers $(n_1, n_2, \ldots, n_N) = n$ the product function $X_n(\omega)$:

$$X_n(\omega) = X_{n_1}(\omega) \cdot X_{n_2}(\omega) \cdots X_{n_N}(\omega)$$

We also use the notation

$$X_\phi(\omega) = 1 \tag{A.4}$$

in which $X_\phi(\omega)$ corresponds to the microcanonical ensemble. One can verify that this set of functions indeed forms an orthonormal basis. This means, as currently used in quantum mechanics (see the section on quantization rules in Chapter 3), that

$$\int_\Omega X_n(\omega)X_{n'}(\omega)\, d\omega = \delta_{n,\,n'}$$

in which $\delta_{n,\,n'}$ is equal to 1 when $n = n'$ (i.e., $n_1 = n'_1, \ldots$ and $n_N = n'_N$) and is otherwise equal to 0. For example, using equation A.3 it is easy to check that (see also Figures A.3 and A.6)

$$\int_\Omega X_1(\omega)X_2(\omega)\, d\omega = 0$$

Also the functions $X_n(\omega)$ together with X_ϕ form a *complete* set: every (square integrable) function on Ω can be expanded in terms of a suitable combination of these functions.

In the following, we shall use the scalar product of two square integrable functions f_1, f_2 defined by

$$\langle f_1, f_2 \rangle = \int_\Omega f_1^{cc}(\omega)f_2(\omega)\, d\omega$$

The baker transformation can also be expressed in terms of an operator U acting on functions $\phi(\omega)$ (as explained in textbooks, see Arnold and Avez,[5] U is a unitary operator):

$$(U\phi)(\omega) = \phi(B\omega) \tag{A.5}$$

As a result we have, using equation A.3,

$$(UX_j)(\omega) = X_j(B\omega)$$
$$= X(u_{j+1})$$
$$= X_{j+1}(\omega) \tag{A.6}$$

Or more generally

$$UX_n = X_{n+1}$$

in which $n + 1$ is the set of integers $(n_1 + 1, n_2 + 1, \ldots, n_N + 1)$. The baker transformation therefore leads to a *simple shift in the basis functions*. We now introduce the characteristic function ϕ_Δ of a domain Δ in Ω. This is the function that takes the value 1 on Δ and vanishes elsewhere in Ω. We can express such characteristic functions in terms of the basis X_n already introduced.

As an example, let us consider $X_1(\omega)$. By definition $X_1(\omega) = X(u_1)$ is a function taking the value -1 when $u_1 = 0$ (i.e., for $0 \leqslant p < \frac{1}{2}$) and $+1$ when $u_1 = 1$ (i.e., for $\frac{1}{2} \leqslant p < 1$). Therefore $X_1(\omega)$ takes the value -1 on the left half of the square, A_1^0, and $+1$ on the right half, A_1^1. It is now easy to write the characteristic functions of the "atoms" of this partition (A_1^0, A_1^1) of the square. The expressions of the characteristic functions are reproduced in Figure A.3. Similar expressions are valid for characteristic functions corresponding to fixed partitions.

We shall now examine the evolution of an arbitrary domain in Ω and link this evolution to the idea of "weak stability" discussed repeatedly in this book (see the section on weak stability in Chapter 2). We may obtain the characteristic function of the transformed domain $B^{-1}\Delta$ from equation A.5, using in succession the definitions

$$\phi_{B^{-1}\Delta}(\omega) = \begin{cases} 1 & \text{if} \quad \omega \in B^{-1}\Delta \\ 0 & \text{if} \quad \omega \notin B^{-1}\Delta \end{cases}$$

$$= \begin{cases} 1 & \text{if} \quad B\omega \in \Delta \\ 0 & \text{if} \quad B\omega \notin \Delta \end{cases}$$

$$= \phi_\Delta(B\omega) = (U\phi_\Delta)(\omega) \tag{A.7}$$

We therefore have

$$\phi_{B^{-1}\Delta} = U\phi_\Delta \tag{A.8}$$

For example, the domain A_1^0 is transformed after n applications of B^{-1} into $B^{-n}A_1^0 = A_{n+1}^0$ (refer to Figure A.3 for the shapes of these domains and their characteristic functions in terms of the basis X_i).

For the general case, we may consider an arbitrary small "atom" of Ω of width $\Delta p = (\frac{1}{2})^n$ and height $\Delta q = (\frac{1}{2})^m$. The set of these $2^n \times 2^m$ "atoms" forms a partition $P_{n,m}$ of Ω. We may write the characteristic functions of such "atoms," $\Delta_{n,m}$, in terms of X_i as follows:

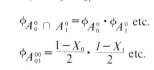

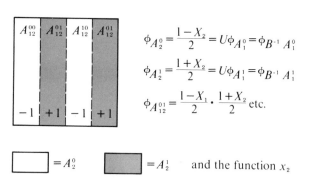

$$x_1 = -1 \quad x_1 = 1$$

$$A_1^0 \quad A_1^1$$

$$\phi_{A_1^0} = \frac{1-X_1}{2}$$

$$\phi_{A_1^1} = \frac{1+X_1}{2}$$

$$A_0^1 \quad x_0 = +1$$

$$A_0^0 \quad x_0 = -1$$

$$\phi_{A_0^0} = \frac{1-X_0}{2}$$

$$\phi_{A_0^1} = \frac{1+X_0}{2}$$

$$A_{01}^{10} \quad A_{01}^{11}$$

$$A_{01}^{00} \quad A_{01}^{01}$$

$$A_0^0 \cap A_1^0 = A_{01}^{00}$$

$$\phi_{A_0^0 \cap A_1^0} = \phi_{A_0^0} \cdot \phi_{A_1^0} \text{ etc.}$$

$$\phi_{A_{01}^{00}} = \frac{1-X_0}{2} \cdot \frac{1-X_1}{2} \text{ etc.}$$

$$A_{12}^{00} \quad A_{12}^{01} \quad A_{12}^{10} \quad A_{12}^{01}$$

$$-1 \quad +1 \quad -1 \quad +1$$

$$\phi_{A_2^0} = \frac{1-X_2}{2} = U\phi_{A_1^0} = \phi_{B^{-1}} A_1^0$$

$$\phi_{A_2^1} = \frac{1+X_2}{2} = U\phi_{A_1^1} = \phi_{B^{-1}} A_1^1$$

$$\phi_{A_{12}^{01}} = \frac{1-X_1}{2} \cdot \frac{1+X_2}{2} \text{ etc.}$$

$$\Box = A_2^0 \qquad \blacksquare = A_2^1 \quad \text{and the function } x_2$$

FIGURE A.3

Examples showing the shapes of the surfaces A_i^j and their intersections $A_{ii'}^{jj'}$ in which the surface A_i^j is the set of points ω such that $u_i = j$ and $A_{ii'\ldots}^{jj'\ldots}$ is the set of points ω such that $u_i = j$, $u_{i'} = j'$, $\ldots$. Moreover, we have given the values of $X_i(\omega)$, $i = 0, 1, 2$, on these surfaces and their characteristic functions $\phi_{A_i^j}$, and so forth.

$$\phi_{\Delta_{n,m}} = \left(\frac{1 \pm X_{-m+1}}{2}\right) \cdot \left(\frac{1 \pm X_{-m+2}}{2}\right) \cdots \left(\frac{1 \pm X_n}{2}\right) \qquad \text{(A.9)}$$

Such atoms can be chosen as small as we want by increasing n and m. The interesting point is that after $(m + 1)$ applications of B^{-1}, $\Delta_{n,m}$ is split into two atoms and this stems from equation A.9:

$$U^{m+1}\phi_{\Delta_{n,m}} = \left(\frac{1 \pm X_2}{2}\right) \cdots \left(\frac{1 \pm X_{n+m+1}}{2}\right)$$

$$= \left(\frac{1 - X_1}{2}\right)\left(\frac{1 \pm X_2}{2}\right) \cdots \left(\frac{1 \pm X_{n+m+1}}{2}\right) \cdot$$

$$+ \left(\frac{1 + X_1}{2}\right)\left(\frac{1 \pm X_2}{2}\right) \cdots \left(\frac{1 \pm X_{n+m+1}}{2}\right)$$

$$= \phi_{\Delta_{0,n+m+1}} + \phi_{\Delta'_{0,n+m+1}} \qquad \text{(A.10)}$$

The two "atoms" so obtained are symmetric and separated by 2^{n+m} subdivisions as shown in Figure A.4. The same result can be obtained by successive applications of B instead of B^{-1}.

We see therefore that, even if at the initial time the system is found in some arbitrary small region of the phase space, it will evolve in time to distinct domains separated in phase space and we can only estimate the probabilities of finding the system in these various domains. In other

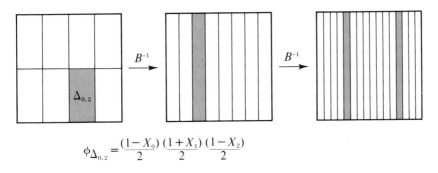

$$\phi_{\Delta_{0,2}} = \frac{(1 - X_0)}{2}\frac{(1 + X_1)}{2}\frac{(1 - X_2)}{2}$$

FIGURE A.4
Splitting of $\Delta_{0,2}$ by two applications of B^{-1}.

words, *each region* (*however small*) contains different types of "trajectories" leading to the various domains. This is the very definition of weak stability.

After these preliminary considerations, let us now introduce the basic operator T corresponding to "age" or to an "internal" time. By definition, it satisfies (for continuous transformations) the "uncertainty relation" (see equation 8.22)

$$-i[L, T] = I$$

The unitary transformation U associated with the discrete baker transformation may be written formally as

$$U = e^{-iL\tau}$$

in which τ corresponds to the time interval between two transformations (we may take $\tau = 1$). The uncertainty relation for L induces the relation

$$[T, U] = U$$

or

$$U^{-1}TU = T + 1 \tag{A.11}$$

In the case of the baker transformation, it is easy to construct the explicit expression of T (for more details, see Misra, Prigogine, and Courbage[1]). We have seen that U when applied to the basis functions $\{X_n\}$ shifts X_n into X_{n+1}.

It is therefore not surprising that the X_n are the eigenvectors of the conjugate operator T. Moreover, for each X_n the corresponding eigenvalue is the maximum of the n_i (recall that n is a finite set of integers $n_1, \ldots n_N$). For example the eigenvalue corresponding to X_n is n, that corresponding to $X_0 X_1 X_2$ is 2, and so forth. As a result, T has the spectral form

$$T = \sum_{n=-\infty}^{+\infty} nE_n \tag{A.12}$$

Here E_n is a projection on the subspace in the orthocomplement of the microcanonical ensemble, generated by the functions X_n: $X_i X_n$ $(i < n)$, $X_i X_j X_n$ $(i, j < n)$, and so forth. We may verify that

$$UE_n U^{-1} = E_{n+1} \qquad \text{(A.13)}$$

and this directly implies relation A.11. The eigenvalues of T (the numerical values of the operator age) are all integers from $-\infty$ till $+\infty$. This has a simple physical meaning. If we consider, for example, an eigenfunction corresponding to age 2, such as X_2, the application of U transforms it into X_3, which is an eigenfunction corresponding to age 3, and so forth.

Not all distributions have a well-defined age: for example, a superposition of X_1 and X_2 has no well-defined age. But we can always introduce an "average age" for a distribution ρ, or more precisely for the excess $\bar{\rho} = \rho - 1$ of ρ with respect to the equilibrium distribution $\rho = 1$. (Note that X_1, being negative in a part of Ω, is not a distribution function, whereas $1 + X_1$, being nonnegative everywhere, is a distribution function.) In conformity with the quantum mechanical definitions, the "average age" of ρ (or $\bar{\rho}$) will be given by

$$T(\rho) = \frac{1}{\|\bar{\rho}\|^2} \langle \bar{\rho}, T\bar{\rho} \rangle \qquad \text{(A.14)}$$

We see, for example, that the "age" of $\rho_\Delta = 1 \pm X_n$ is well defined:

$$T(\rho_\Delta) = \langle X_n, TX_n \rangle = n \qquad \text{(A.15)}$$

More generally, we may write $\bar{\rho}$ in the form

$$\bar{\rho} = \sum_n C_n X_n = \sum_{n=-\infty}^{+\infty} \left(\sum_{n(n)} C_n X_n \right) \qquad \text{(A.16)}$$

in which $n(n)$ is the set of integers having the maximum n. So from equation A.14 we get

$$T(\rho) = \frac{1}{\|\bar{\rho}\|^2} \sum_n n \left\langle \sum_{n'(n')} C_{n'} X_{n'}, \sum_{n(n)} C_n X_n \right\rangle$$

$$= \sum_{n=-\infty}^{+\infty} n \left(\frac{1}{\|\bar{\rho}\|^2} \sum_{n(n)} |C_n|^2 \right)$$

$$= \sum_{n=-\infty}^{+\infty} n\mu_n \qquad \text{(A.17)}$$

in which the μ_n are positive coefficients:

$$\mu_n = \frac{1}{\|\bar{\rho}\|^2} \sum_{n(n)} |C_n|^2 \tag{A.18}$$

The coefficient μ_n represents the probability of finding the system at age n. Indeed, we may speak of probability because we have:

$$\sum_{n=-\infty}^{\infty} \mu_n = 1 \tag{A.19}$$

The similarity with the rules of quantum mechanics is striking. However, there is a simple explanation for this: the uncertainty relation (equation A.11) between the time evolution and the age T. We may also define the fluctuation in age and other characteristics of the stable distribution of ages.

The microscopic entropy operator M follows directly. It is the sum of two terms. One is closely related to the eigenfunctions of T. The corresponding eigenprojectors are E_n and its eigenvalues form a decreasing sequence λ_n^2 of real numbers tending to zero for $n \rightarrow \infty$ and to 1 for $n \rightarrow -\infty$.

In addition, M contains the projection operator on the microcanonical ensemble:

$$P_0\rho = \langle \rho, 1 \rangle \cdot 1 = \int_\Omega \rho(\omega)\, d\omega \tag{A.20}$$

We may therefore write M in the form

$$M = \sum_{n=-\infty}^{+\infty} \lambda_n^2 E_n + P_0 \tag{A.21}$$

There is a whole set of entropy operators according to the choice of the sequence $\{\lambda_n^2\}$. Let us verify that definition A.21 indeed leads to the correct behavior of the Lyapounov function $\Omega(\rho)$ defined as

$$\Omega(\rho) = \langle \rho, M\rho \rangle \tag{A.22}$$

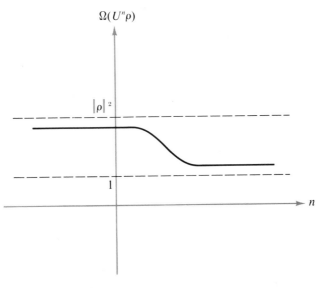

FIGURE A.5
The behavior of $\Omega(U^n\rho)$ for a normalized state ρ.

Indeed, using equations A.13 and A.22, we have

$$\Omega(U\rho) - \Omega(\rho) = \langle U\rho, MU\rho \rangle - \langle \rho, M\rho \rangle$$

$$= \sum_{n=-\infty}^{\infty} (\lambda_{n-1}^2 - \lambda_n^2)\|E_n\rho\|^2 \qquad (A.23)$$

The basic property is the monotonous time variation of Ω (it is a matter of definition to require that it *increases* or *decreases* monotonously with time). Each step decreases the value of the Lyapounov variable (Figure A.5). The monotonous variation of Ω is the consequence of the existence of the time operator, T, which itself results from the "weak stability" of the dynamic transformation. No approach to probabilistic considerations is necessary.

Several important results and properties follow. First, the nonunitary transformation Λ introduced in the text can be explicitly constructed. It corresponds to the square root of M and has the spectral form

$$\Lambda = \sum_{n=-\infty}^{+\infty} \lambda_n E_n + P_0$$

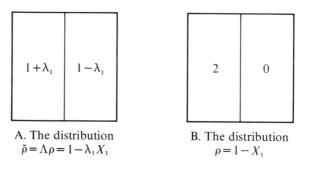

A. The distribution
$\tilde{\rho} = \Lambda\rho = 1 - \lambda_1 X_1$

B. The distribution
$\rho = 1 - X_1$

FIGURE A.6
The "delocalization" of $\rho = 1 - X_1$.

This transformation induces a contraction semigroup corresponding to the generator $\tilde{U} = \Lambda U \Lambda^{-1}$. Indeed, we now have a monotonously decreasing norm (see equation A.23):

$$\|\tilde{U}^n\tilde{\rho}\|^2 = \langle \Lambda U^n\rho, \Lambda U^n\rho \rangle$$

$$= \langle U^n\rho, M U^n\rho \rangle$$

in which $\tilde{\rho} = \Lambda\rho$. It can be shown that this transformation takes a state into a state (it preserves the positivity and the normality). Another important property of this transformation is that it delocalizes the distribution. This means that, if ρ has a nonvanishing value only in a region Δ, then $\Lambda\rho$ has a nonvanishing density almost everywhere. In the Λ-picture, the approach to equilibrium corresponds to the cancellation of the local excess with respect to the equilibrium in each region. We can see this property in the simple example of the distribution $\rho = 1 - X_1$ (Figure A.6). This distribution has a value of 2 on the left half of the square and vanishes on the right, whereas $\Lambda\rho = 1 - \lambda_1 X_1$ is positive everywhere because λ_1 is strictly positive and less than one. We check that

$$\tilde{U}^n\tilde{\rho} = 1 - \tilde{X}_{1+n}$$

with

$$\tilde{X}_{1+n} = \Lambda X_{1+n} = \lambda_{1+n} X_{1+n}$$

X_{1+n} is the excess with respect to the equilibrium distribution and tends to zero with increasing n, for we have

$$\|\tilde{X}_{1+n}\|^2 = \lambda^2_{1+n} \xrightarrow[n \to \infty]{} 0$$

Other concepts could also be introduced, such as "measures" associated with partitions of the square, but such details will not be dealt with here.

This simple example shows how geometry, dynamics, and irreversibility can be linked in the framework of an extension of classical dynamics involving supplementary observables such as T or M represented by suitable operators.

We also see how the requirement of a universal Lyapounov function can be satisfied using nonunitary transformations (the "square root" of M), which themselves depend on the dynamics of the system. The analogy with the basic ideas of general relativity (the use of geometrical concepts to express the laws of dynamics in a simple way) is striking.

References

1. B. Misra, I. Prigogine, and M. Courbage, *Proceedings of the National Academy of Sciences, U.S.A.* 76 (1979): 3607; *Physica* 98A (1979): 1.
2. J. L. Lebowitz, *Proceedings of I.U.P.A.P. Conference on Statistical Mechanics* (Chicago, 1971).
3. D. S. Ornstein, *Advances in Mathematics*, 4 (1970): 337.
4. J. G. Sinai, *Theory of Dynamical Systems*, vol. 1 (Denmark: Aarhus University, 1970).
5. V. I. Arnold, and A. Avez, *Ergodic Problems of Classical Mechanics* (New York: Benjamin, 1968).
6. P. Shields, *The Theory of Bernouilli Shifts* (Chicago: University of Chicago Press, 1973).

RESONANCES AND KINETIC DESCRIPTION

As noted in Chapter 8 (see the sections titled Construction of the Entropy Operator and the Transformation Theory and Entropy Operator and the Poincaré Catastrophe), the microscopic operator M can exist in two cases: one corresponds to systems with strong mixing properties (an example is given in Appendix A); the other corresponds to the "Poincaré catastrophe." In the second case, an important role is played by the collision operator, $\Psi(z)$, defined in 8.33. In this appendix, a case will be presented in which the collision operator can be calculated explicitly. Although the example to be considered is somewhat schematic, it shows the meaning of resonances, so important in Poincaré's approach (see Chapter 3), and their relation to the limit of Ψ as $z \to 0$, which corresponds to long-time behavior and which is the basic quantity appearing in kinetic theory. To see this more clearly, the section on the entropy operator and the Poincaré catastrophe in Chapter 8 must be developed further.

The decomposition of the resolvent of L can be written in terms of the resolvent of QLQ:

$$(L - z)^{-1} = [P + \mathscr{C}(z)][\Psi(z) - z]^{-1}[P + \mathscr{D}(z)] + \mathscr{P}(z) \qquad \text{(B.1)}$$

in which

$$\mathscr{C}(z) = -(QLQ - z)^{-1}QLP \qquad \text{(B.2)}$$

$$\mathscr{D}(z) = -PLQ(QLQ - z)^{-1} \qquad \text{(B.3)}$$

$$\mathscr{P}(z) = -Q(QLQ - z)^{-1} \qquad \text{(B.4)}$$

$$\Psi(z) = PLP + \Psi(z) \qquad \text{(B.5)}$$

with Ψ the collision operator, defined in 8.33. The creation operator, $\mathscr{C}(z)$, "creates" correlations (vectors in the Q-subspace) out of the vacuum (vectors in the P-subspace). Similarly, $\mathscr{D}$ and $\mathscr{P}$ are the destruction and propagation operators.[1] Laplace inversion of equation B.1 gives the solution of the Liouville equation in the time variable:

$$|\rho(t)\rangle = \frac{-1}{2\pi i} \int_B dz e^{-itz}\{[P + \mathscr{C}(z)]$$

$$\times [\Psi(z) - z]^{-1} \cdot [P + \mathscr{D}(z)] + \mathscr{P}(z)\}|\rho(0)\rangle \qquad \text{(B.6)}$$

in which $\langle J\alpha|\rho(t)\rangle$ is the density-in-phase and B is a line parallel to the real z axis, directed from right to left, and situated above all singularities of the integrand. The P projection of equation B.6 obeys the exact non-Markovian master equation, obtained by Résibois and myself some years ago:[1, 2]

$$\frac{\partial}{\partial t} P|\rho\rangle = -iPLP \cdot P|\rho\rangle - \int_0^t d\tau G(t - \tau)P|\rho(\tau)\rangle$$

$$+ F(t)Q|\rho(0)\rangle \qquad \text{(B.7)}$$

in which G and F are the inverse Laplace transforms of Ψ and $\mathscr{D}$. As can be seen in equation B.7, the evolution of $P|\rho(t)\rangle$ depends on the values of $P|\rho(t)\rangle$ at all earlier times as well as on the initial correlations. However, in the long-time limit $t \to \infty$, equation B.7 can, under suitable conditions, be replaced by the following Markovian kinetic equation, obtained by Résibois and myself:

$$i \frac{\partial}{\partial t} P |\rho(t)\rangle = \Omega\Psi(i0+)P|\rho(t)\rangle \tag{B.8}$$

which is equation 8.34 corrected by the operator Ω.

The nature of $\Psi(i0+)$ and of $\mathscr{C}(i0+)$ places restrictions on the properties of the dynamical invariants, as was seen in equation 8.35. Further, by pointing out that

$$\lim_{t \to \infty} \frac{1}{t} \int_0^t d\tau \exp(-i\tau L) = E = \lim_{z \to i0+} z(z - L)^{-1} \tag{B.9}$$

in which E is the projector onto the null space of L (the subspace of the dynamical invariant), Stey showed how the invariants, the infinite-time averages, and the value of the limit of Ψ as $z \to i0+$ are related.[3, 4] He showed that, when $[z - \Psi(z)]^{-1}$ possesses a simple pole as singularity at $z = 0$, and

$$P|g\rangle = \lim_{t \to \infty} \frac{1}{t} \int_0^t d\tau P e^{-i\tau L} P |f\rangle \tag{B.10}$$

then to $P|g\rangle$ there corresponds only one $P|f\rangle$ if and only if $\Psi(z) \to 0$ as $z \to i0+$. Figure B.1 is a schematic representation of the relation between the initial value of the above time-average, $P|f\rangle$, and its final value $P|g\rangle$, as a function of $\Psi(i0+)$. Equivalent to this result is $\Psi(i0+) = 0$ if and only if zero is the only P-subspace vector orthogonal to each invariant.[3]

Consider the example studied by Stey[3] that was mentioned earlier. The system has one degree of freedom and the state (J, α) in action-angle variables evolves according to Hamilton's equations:

$$\frac{dJ}{dt} = -\frac{\partial H}{\partial \alpha}; \quad \frac{d\alpha}{dt} = \frac{\partial H}{\partial J} \tag{B.11}$$

and the Hamiltonian is taken as

$$H(J, \alpha) = \omega J - 2V \sin(\alpha) \tag{B.12}$$

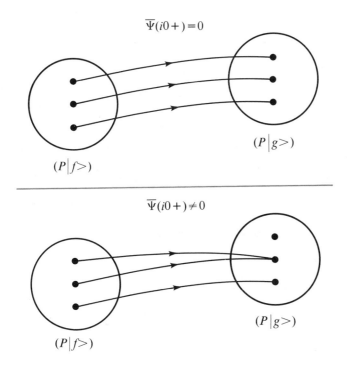

$\overline{\Psi}(i0+)=0$

$(P|f>)$

$(P|g>)$

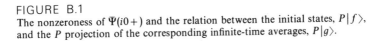

$\overline{\Psi}(i0+)\neq0$

$(P|f>)$

$(P|g>)$

FIGURE B.1
The nonzeroness of $\Psi(i0+)$ and the relation between the initial states, $P|f\rangle$, and the P projection of the corresponding infinite-time averages, $P|g\rangle$.

From the solution of equation B.11, we see that two types of flow are possible. First, when $\omega \neq 0$, each bounded region of phase fluid in the (J, α) plane undergoes a periodic deformation while performing an oscillatory motion in a band parallel to the α axis (see Figure B.2). The second kind of motion appears in the limit $\omega \to 0$, in which each region flows parallel to the J axis and stretches itself out continuously along the line of flow, becoming infinitely long as $t \to \infty$ (see Figure B.3). In the limit $\omega \to 0$, the distance between points in the (J, α) plane diverges at t when $t \to \infty$.

To analyse this system from the point of view of the Liouville equation (2.12), we first look at the matrix elements of the Liouville operator (2.13). By definition of L, we have

$$\langle J\alpha|L|f\rangle = -i\left(\frac{\partial H}{\partial J}\frac{\partial}{\partial \alpha}\langle J\alpha|f\rangle - \frac{\partial H}{\partial \alpha}\frac{\partial}{\partial J}\langle J\alpha|f\rangle\right) \qquad (B.13)$$

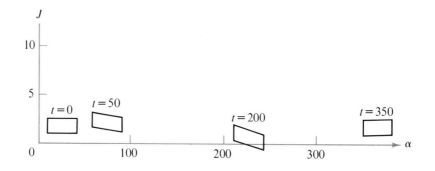

FIGURE B.2
Oscillatory flow in phase space ($\omega \neq 0$).

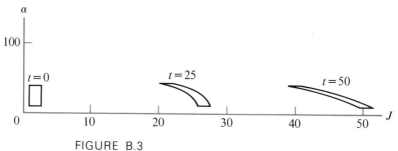

FIGURE B.3
Nonoscillatory unbounded flow in phase space ($\omega = 0$).

and

$$\langle J_1 \alpha_1 | L | J_2 \alpha_2 \rangle = -i[\omega \, \delta'(\alpha_1 - \alpha_2) \, \delta(J_1 - J_2)$$

$$+ 2V \cos(\alpha_1) \, \delta(\alpha_1 - \alpha_2) \, \delta'(J_1 - J_2)] \quad \text{(B.14)}$$

By straightforward diagonalization of L, we obtain (see Stey[3]) the matrix elements of the resolvent of L; for $\omega \neq 0$,

$$\langle J_1 \alpha_1 | (L - z)^{-1} | J_2 \alpha_2 \rangle$$

$$= \delta \left[J_1 - J_2 - \frac{2V}{\omega} (\sin \alpha_1 - \sin \alpha_2) \right]$$

$$\cdot \frac{1}{2\pi} \sum_{n=-\infty}^{+\infty} \frac{e^{in \cdot (\alpha_1 - \alpha_2)}}{(n\omega - z)} \quad \text{(B.15)}$$

The delta function in equation B.15 expresses the restriction of transitions to those on the energy surface. Laplace inversion of equation B.15 gives:

$$\langle J_1\alpha_1 | e^{-itL} | J_2\alpha_2 \rangle = \delta\left[J_1 - J_2 - \frac{2V}{\omega}(\sin\alpha_1 - \sin\alpha_2) \right]$$

$$\cdot \sum_{l=-\infty}^{+\infty} \delta(\alpha_1 - \alpha_2 - \omega t + 2\pi l) \qquad (B.16)$$

The series of delta functions in equation B.16 describes the motion of the angle variable in phase space.

To study the distribution of J values over a statistical ensemble, we introduce the P, Q decomposition of equations 8.29 and B.1 and define P through its matrix elements:

$$\langle J_1\alpha_1 | P | J_2\alpha_2 \rangle = \frac{\delta(J_1 - J_2)}{2\pi} \qquad (B.17)$$

$\langle J_1\alpha_1 | P | \rho(t) \rangle$ thus comprises the density-in-action variables. An explicit evaluation of the collision operator in this case has been given:[3] $PLP = 0$ here, $\bar{\Psi} = \Psi$, and

$$\langle J_1\alpha_1 | \Psi(z) | J_2\alpha_2 \rangle = \frac{1}{4\pi^2} \int_{-\infty}^{+\infty} d\kappa\, e^{i\kappa(J_1 - J_2)} \Psi_\kappa(z) \qquad (B.18)$$

in which

$$\Psi_\kappa(z) = z - \left[\sum_{v=-\infty}^{+\infty} \mathscr{J}_v^2(\beta\kappa)/(z - \mu\omega) \right]^{-1}$$

$$= z - \left\{ \frac{1}{i}(1 - e^{(2\pi i/\omega)z})^{-1} \int_0^{2\pi/\omega} dt\, e^{itz} \mathscr{J}_0\left[2\beta\kappa \sin\left(\frac{\omega t}{2}\right) \right] \right\}^{-1}$$

$$(B.19)$$

in which $\beta = 2V/\omega$ and $\mathscr{J}_v(x)$ is the Bessel function of the first kind. Looking at the small z behavior of the collision operator eigenvalues, we obtain

$$\Psi_\kappa(z) \sim [1 - \mathscr{J}_0^{-2}(\beta\kappa)]z \to 0 \qquad (B.20)$$

as $z \to i0+$ and $\omega \neq 0$, provided that $\beta\kappa$ is not equal to the zero of $\mathscr{J}_0(x)$. The Fourier transform $\tilde{g}(\kappa)$ of the infinite-time average of a state $\langle J\alpha|P|\rho(t)\rangle$ whose initial value has Fourier transform $\tilde{f}(\kappa)$ is, from equations 8.32, B.18, and B.19, given by

$$\tilde{g}(\kappa) = \mathscr{J}_0^2(\beta\kappa)\tilde{f}(\kappa) \tag{B.21}$$

which is equation B.10 in this case. It can be seen by explicit calculation of the resolvent of QLQ that the conditions associated with the results of equation B.10 are satisfied here. Indeed, equation B.21 shows immediately that there is one-to-one correspondence between $\tilde{g}(\kappa)$ and $\tilde{f}(\kappa)$ if and only if $\mathscr{J}_0(\beta\kappa) \neq 0$, which is so if and only if $\Psi(i0+) = 0$, as explicitly seen above. The solution of equation B.7 can be obtained by inverse Laplace transformation, once $\Psi(z), [z - \Psi(z)]^{-1}$, and the resolvent of QLQ are computed. It can be written in the form

$$\langle J_1\alpha_1|P|\rho(t)\rangle = \frac{1}{2\pi}\int_0^{2\pi} d\alpha_2$$

$$h\left(J_1 - \frac{2V}{\omega}[\sin\alpha_2 - \sin(\alpha_2 - \omega t)], \alpha_2 - \omega t\right) \tag{B.22}$$

in which $\langle J_1\alpha_1|P|\rho(0)\rangle = h(J_1, \alpha_1)$ and h is 2π-periodic in α_1. Thus, we see that, when $\omega \neq 0$ and $\Psi(i0+) = 0$, the above solution is T-periodic in time, in which $T = 2\pi/\omega$, the Poincaré recurrence time. On the other hand, in the limit $\omega \to 0$, the Poincaré recurrence time is pushed to infinity and equation B.19 shows that

$$\lim_{\omega \to 0} \Psi_\kappa(z) = \frac{-i|2V\kappa|^2}{(|2V\kappa|^2 - z^2)^{1/2} - iz}; \quad \text{im}(z) > 0 \tag{B.23}$$

and so

$$\lim_{z \to i0+}\lim_{\omega \to 0} \Psi_\kappa(z) = -2i|V\kappa| \tag{B.24}$$

Furthermore, when $\omega \to 0$, equation B.21 becomes simply

$$\tilde{g}(\kappa) = 0 \tag{B.25}$$

Thus, the one-to-one correspondence between $\tilde{f}(\kappa)$ and $\tilde{g}(\kappa)$ mentioned earlier no longer holds: for all suitable initial states the infinite time average is the same—zero. This is of course in complete agreement with the nonzero value of the asymptotic collision operator of equation B.24. Moreover, in the $\omega \to 0$ limit equation B.22 tends to

$$\langle J_1\alpha_1 | P | \rho(t) \rangle = \frac{1}{2\pi} \int_0^{2\pi} d\alpha_2 \, h(J_1 - 2tV \cos(\alpha_2), \alpha_2) \qquad \text{(B.26)}$$

Upper bounds on the rate of relaxation, which is now possible, have been obtained by application of the Hölder inequality. For all $q > 2$ and $\omega = 0$,

$$|\langle J_1\alpha_1 | P | \rho(t) \rangle| \leqslant \frac{C}{|2Vt|^{1/q}} \left\{ \int_{-\infty}^{+\infty} dJ \left[\sup_{0 \leqslant \alpha \leqslant 2\pi} |h(J, \alpha)| \right]^q \right\}^{1/q} \qquad \text{(B.27)}$$

in which C is a constant. This result suggests that we recall that an important point concerning the flow of a region of phase fluid in this model is that each region must return to its initial shape after a period, $2\pi/\omega$. When $\omega = 0$, each region, no matter how small, of nonzero area becomes *infinitely extended* parallel to the J axis as $t \to \infty$. Thus, we see that when $\Psi(i0+) \neq 0$ here, the idea of a deterministic trajectory in phase space can be useful operationally in the long-time limit *only* when the initial state of the system is known *exactly*. In this connection, the description by kinetic equation therefore becomes of special interest. The analogy with mixing systems should also be stressed. A comparison of the solution of the kinetic equation (B.8) with the exact long-time decay of the right-hand member of equation B.26 has been made. Taking

$$h(J_1\alpha_1) = \frac{\delta(J_1 - J_2)}{2\pi} \qquad \text{(B.28)}$$

and limiting ourselves to the simplest approximation, $\Omega = 1$ for Ω in equation B.8, we have equation 8.34, which, using the value given in equation B.24, gives[3]

$$\langle J_1\alpha_1 | P e^{-it\Psi(i0+)} | \rho(0) \rangle = \frac{1}{|2\pi Vt|} \cdot \left[1 + \left(\frac{J_1 - J_2}{2Vt} \right)^2 \right]^{-1} \qquad \text{(B.29)}$$

From equation B.26, we have the following result:

$$\langle J_1\alpha_1 | Pe^{-itL} | \rho(0)\rangle = \frac{1}{|2\pi Vt|} \cdot \left[1 - \left(\frac{J_1 - J_2}{2Vt}\right)^2\right]^{-1/2} \tag{B.30}$$

We see that equations B.29 and B.30 are in good agreement for $t \to \infty$ and that the initial distribution, although infinitely peaked as a function of J, leads to a uniform distribution of action as $t \to \infty$. This does not occur when the initial α distribution is infinitely peaked about a certain value: that is, when a trajectory is considered.

References

1. I. Prigogine, *Non-equilibrium Statistical Mechanics* (New York: Interscience, 1962); I. Prigogine, C. George, F. Henin, and L. Rosenfeld, *Chem. Scripta* 4 (1973): 5.
2. I. Prigogine and P. Résibois, *Physica* 27 (1961): 629; I. Prigogine and A. Grecos, *Problems in the Foundations of Physics* (Varenna, Italy: International School of Physics "Enrico Fermi," 1977).
3. G. C. Stey, *Physics Letters* 69A (1978): 151.
4. I. Prigogine, A. Grecos, and C. George, *Proceedings of the National Academy of Sciences* 73 (1976): 1802.

Appendix C

ENTROPY, MEASUREMENT, AND THE SUPERPOSITION PRINCIPLE IN QUANTUM MECHANICS

Pure States and Mixtures

As noted in Chapter 3, a fundamental distinction is made in quantum mechanics between pure states (wave functions) and mixtures represented by density matrices. Pure states occupy a privileged position in quantum mechanics, somewhat analogous to orbits in classical mechanics. As indicated by the Schrödinger equation (see equations 3.17 and 3.18), pure states are transformed into other pure states during the time evolution. Moreover, observables are defined as Hermitian operators mapping vectors of the Hilbert space into itself. These operators also preserve pure states. The basic laws of quantum mechanics can thus be formulated without ever invoking the density-matrix description of states corresponding to mixtures. The use of that description is considered to be only a matter of practical convenience or approximation. The situation is similar to that considered in classical dynamics in which the basic element corresponding to the pure state is the orbit or the trajectory of a dynamical system (see, in particular, Chapters 2 and 7).

In Chapter 3, the question was asked: Is quantum mechanics complete? We have seen that one of the reasons for asking this question in spite of the striking successes of quantum mechanics in the past fifty years is the difficulty of incorporating the measurement process (see the section titled The Measurement Problem in Chapter 3). We have seen that the measurement process transforms a pure state into a mixture and therefore cannot be described by the Schrödinger equation, which transforms a pure state into another pure state.

In spite of much discussion (see the beautiful account by d'Espagnat[1]), this problem is far from being solved. According to d'Espagnat (p. 161), "The problem [of measurement] is considered as non-existent or trivial by an impressive body of theoretical physicists and as presenting almost insurmountable difficulties by a somewhat lesser but steadily growing number of their colleagues."

I do not wish to take a position that is too strong in this controversy, because, for the present purpose, the measurement process is simply an illustration of the problem of irreversibility in quantum mechanics.

Whatever the position one takes, the fundamental distinction between pure states and mixtures and the privileged position of the pure states in the theory must be given up. Thus, the problem is to provide a fundamental justification for this loss of distinction. It is a remarkable fact that the introduction of the entropy operator M (see the section on irreversibility and the formalism of classical and quantum mechanics in Chapter 8) as a fundamental object of the theory entails just this loss of distinction between pure states and mixtures.

The object of this appendix is to sketch a proof of this statement. For more details, the reader is referred to a soon-to-be published paper by Misra, Courbage, and myself[2] on which the present appendix is based.

Entropy Operator and Generator of Motion

Why must we go beyond the standard formulation of quantum mechanics in which the Hamiltonian operator is the generator of motion according to the Schrödinger equation (3.17)? Suppose that we have (see equation 7.27, for convenience the sign of D has been changed)

$$i[H, M] \equiv D \geqslant 0 \qquad (C.1)$$

We can then view D as the microscopic entropy production operator. It seems natural to suppose that the measurements of M and D are mutually compatible. As is well known, this implies

$$[M, D] = 0 \qquad (C.2)$$

Equation C.2 can be considered a "sufficient condition" for all that follows. It could be weakened, but it is not necessary to go into greater detail here.

The basic reason that conditions C.1 and C.2 cannot be satisfied by an operator M is that the Hamiltonian operator H plays a dual role in quantum mechanics (see the section on particles and dissipation in Chapter 8). In addition to generating the time evolution, it represents the energy of the system. Hence it must be bounded from below.

$$H \geqslant 0 \qquad (C.3)$$

To see the incompatibility of the positivity of the Hamiltonian H with conditions C.1 and C.2, consider the identity

$$\frac{d}{dt} \langle e^{-iMt}\psi, He^{-iMt}\psi \rangle = -i\langle e^{-iMt}\psi, [H, M]e^{-iMt}\psi \rangle$$

$$= -\langle \psi, D\psi \rangle \qquad (C.4)$$

The last equality follows from the fact that M and D commute so that $e^{iMt}De^{-iMt} = D$. Integration of both sides of equation C.4 from 0 to t now yields

$$\langle e^{-iMt}\psi, He^{-iMt}\psi \rangle - \langle \psi, H\psi \rangle = -t\langle \psi, D\psi \rangle$$

or

$$\langle \psi, H\psi \rangle = t\langle \psi, D\psi \rangle + \langle e^{-iMt}\psi, He^{-iMt}\psi \rangle \qquad (C.5)$$

because

$$H \geqslant 0, \qquad \langle \psi, H\psi \rangle \geqslant t\langle \psi, D\psi \rangle$$

for all t. But this is clearly impossible except in the trivial case in which $D = 0$.

There is an interesting connection between the nonexistence of the entropy operator M and the impossibility (noted by Pauli[3]) of defining an operator of time T in the usual formulation of quantum mechanics. Such an operator of time would be canonically conjugate with the generator H of the time evolution group—or (see also the section titled Is Quantum Mechanics Complete? in Chapter 3 and that titled The Construction of the Entropy Operator and the Transformation Theory, in Chapter 8)

$$i[H, T] = I \qquad (C.6)$$

However, if a self-adjoint operator T satisfying equation C.6 exists, an entropy operator M satisfying equations C.1 and C.2 can be obtained by simply taking M to be a monotonic function of T:

$$M = f(T)$$

The impossibility of defining the entropy operator M, the nonexistence of a time operator in quantum mechanics, and the problem of interpreting and justifying the time-energy uncertainty relation are thus linked together. Their common origin is the fact that in the usual formulation of quantum mechanics the generator H of the time-translation group is identical with the energy operator of the system. To be able to define the entropy operator M, it is thus necessary to overcome this degeneracy. The simplest way of achieving this is to go to the so-called Liouvillian formulation of (quantum) dynamics (see the section titled Shrödinger and Heisenberg Representations in Chapter 3). The basic object in this formulation is the group describing the time evolution of the density operators. As noted in Chapter 3, the generator of the time-translation group is now the Liouvillian operator L defined by the equation (see equations 3.35 and 3.36)

$$L\rho = [H, \rho] \qquad (C.7)$$

Let us therefore investigate the existence of M in conjunction with the time evolution as generated by the Liouville operator.

The Entropy Superoperator

An important advantage of using the Liouvillian formulation of quantum dynamics is that the generator L of the time-translation group is no longer physically constrained to be bounded from below. In fact, if the spectrum of H extends from 0 to $+\infty$, the spectrum of L is the entire real line. The possibility of defining M as a "superoperator" (see the section titled Irreversibility and the Formalism of Classical and Quantum Mechanics in Chapter 8) that satisfies the relations

$$i[L, M] = D \geqslant 0$$

and

$$[M, D] = 0 \tag{C.8}$$

is thus not excluded by the argument given earlier.

As in classical mechanics, supplementary conditions (see the section on ergodic systems in Chapter 2) must be imposed; M cannot exist in both of the following cases:

1. H has a purely discrete spectrum.
2. H has a continuous but bounded spectrum.

In physical terms, this means that the entropy superoperator cannot exist for a finitely extended system consisting of only a finite number of particles. This clearly shows that the question is much more involved than that for classical mechanics—as in Appendix A, M has been constructed for a classical finite system involving only a single degree of freedom.

A fundamentally important property of the entropy superoperator M is that it is necessarily *nonfactorizable*. This means that $M\rho$ is *not* of the form

$$M\rho = A_1 \rho A_2 \tag{C.9}$$

in which A_1 and A_2 are usually self-adjoint operators. A first remark is that, if M were factorizable, it could be written in the simpler form

$$M\rho = A\rho A \qquad (C.10)$$

using general properties such as preservation of Hermiticity (see note 2). Such a factorizable operator would preserve pure states. It would in fact simply transform $|\psi\rangle$ into $A|\psi\rangle$ (see equation 3.30).

The nonfactorizability of M is therefore a very important property. In fact, if $M\rho$ is given by equation C.10, it is not difficult (see note 2) to verify that the commutation relations (C.8) for M lead to the following relation for A:

$$i[H, A] = D_1 \geqslant 0$$

$$i[D_1, A] = cA^2 \qquad (C.11)$$

in which c is a real number. The three cases that arise (corresponding to $c = 0$, $c > 0$ and $c < 0$) can be ruled out separately as follows:

1. $c = 0$. The preceding argument now shows that $[H, A] = D_1 = 0$. This together with equation C.10 then leads to the relation

$$i[L, M] \equiv D \equiv 0 \qquad (C.12)$$

which means that M is an invariant of motion.

2. $c > 0$. In this case,

$$D_1 \geqslant 0$$

$$i[D_1, A] = cA^2 \qquad (C.13)$$

This case can be studied by means of a formal analogy with the second section of this appendix; the positive operator D_1 plays the role of H; A the role of M; and cA^2 of D. We may therefore directly conclude that $A^2 = 0$ and hence M as given by equation C.10.

The preceding considerations thus lead us to the following conclusions. For infinite quantal systems, there exists the possibility of enlarging the algebra of observables to include an operator M representing nonequilibrium entropy. The operator M can be defined, however, only as a non-factorizable superoperator. The inclusion of an entropy operator (necessarily nonfactorizable) among the observables thus requires that the pure states lose their privileged position in theory and that the pure and mixed states be treated on an equal basis. Physically, this means that, for systems having entropy as an observable, the distinction between pure and mixed states must cease to be operationally meaningful and there would be limitations on the possibility of realizing coherent superposition of quantum states (see the section titled The Measurement Problem in Chapter 3).

Evidently, this conclusion, which has been reached as a *logical* consequence of our theory of entropy operator, should be further elucidated by an analysis of the physical reason for the loss of the distinction between pure and mixed states.

The situation for classical systems has been discussed repeatedly (see Chapters 3, 7, 8, and 9). We have seen that there are at present two mechanisms known to lead to instability of motion, which in turn prohibits the "observation" of well-defined trajectories. It can be expected that the physical reason for the loss of the distinction between pure and mixed states of quantal systems with an entropy operator is some suitable quantum analogue of the instability discussed earlier.

Again, as in classical mechanics, there may be more than one mechanism involved. One may be the existence of analogues of classical systems with strong mixing properties; the other existence of the Poincaré catastrophe for quantum systems (see the section on the entropy operator and the Poincaré catastrophe in Chapter 8). One simple example for classical systems in which the asymptotic collision operator $\Psi(z)$ for $z \to 0$ does not vanish was discussed in Appendix B. Similar situations exist in the quantum case and play an essential role in the derivation of kinetic equations (see equation 8.34). A rigorous mathematical formulation of the quantum instability mechanism is work for the future. Nevertheless, it is satisfying that the second principle of thermodynamics when interpreted as a dynamical principle in terms of the existence of the operator M requires us to give up the distinction between pure and mixed states in precisely the situation in which this distinction is expected to be physically unobservable.

References

1. B. d'Espagnat, *Conceptual Foundations of Quantum Mechanics*, 2d ed. (Menlo Park, California: Benjamin, 1976).
2. B. Misra, I. Prigogine, and M. Courbage, *Proceedings of the National Academy of Sciences*, in press.
3. See M. Jammer, *The Philosophy of Quantum Mechanics* (New York: Wiley-Interscience, 1974), p. 141.

COHERENCE AND RANDOMNESS
IN QUANTUM THEORY

Operators and Superoperators

In Chapter 9, emphasis was on the important role of instability in the foundations of statistical physics. Appendix A demonstrated the possibility of obtaining stochastic (Markov) processes, starting with deterministic dynamics, through an appropriate nonunitary "change of representation" that does not entail any loss of information. It is possible to define this change of representation, provided the dynamics of the system has a suitable high degree of instability. This proves that a probabilistic theory can still be "complete" and "objective."

The viewpoint adopted in Appendix A is that, when a (classical) dynamical system is sufficiently unstable, we can no longer speak of trajectories and we are forced to deal with a basically different approach: the evolution of distribution functions (or bundles of trajectories) in phase space. The transition from distribution functions to a single point in phase space cannot be performed under such conditions (see the section on time and change in Chapter 9).

In quantum theory, coordinates and momenta retain their meaning, and measurements can delimitate in the appropriate phase space a region in which the system is located.

The question may then be asked: Are there other circumstances, such as those related to the formulation of quantum theory, in which the transition from phase-space distribution functions to individual trajectories is also impossible?

Usually another attitude is adopted: the transition to a single trajectory is performed before the question of the relation between classical and quantum mechanics is raised. However, the concepts of a classical trajectory and of a quantum wave function are so different that it is difficult to compare them in a meaningful way.

The type of problem met with here is quite different from that encountered in classical theory. There we deal with unstable, "disordered" systems—in fact, so disordered that it becomes possible to define Lyapounov functions closely related to entropy. In contrast, the transition from classical to quantum mechanics does not affect the basic reversibility of classical dynamics (see Chapter 3). Moreover, as mentioned in Chapter 3 in the section on the decay of unstable particles, all finite systems of quantum mechanics have a discrete energy spectrum and therefore a purely periodic motion. Quantum theory leads in this sense to a more "coherent" behavior of motion than does classical theory. This can be considered a strong physical argument against any attempt to understand quantum theory in terms of "hidden" variables or in terms of traditional stochastic models. On the contrary, this increased coherence seems to indicate that quantum theory should correspond to an "overdetermined" classical theory. Alternatively, it seems that quantum effects lead to correlations between neighboring classical trajectories in phase space. This is what the old Bohr-Sommerfeld image of phase cells of area h expresses in an intuitive way.

To express this idea in a new, precise way,[1] we must reintroduce a basic distinction between operators and superoperators.[2] This distinction was already discussed in Chapter 8 in the section titled Irreversibility and the Formalism of Classical and Quantum Mechanics. It has also been mentioned (equation 3.35 and Appendix C) that the Liouville operator is a factorizable superoperator. By definition, a factorizable superoperator F may be written as $A_1 \times A_2$ (see equation C.9) with the meaning

$$F\rho = (A_1 \times A_2)\rho = A_1\rho A_2 \qquad (D.1)$$

Using this notation, we have for the Liouville superoperator

$$L = \frac{1}{h}(H \times I - I \times H) \qquad \text{(D.2)}$$

The factorizability of quantum superoperators is a fundamental property, without classical analogue. For example the classical Liouville operator L_{cl} is also a *superoperator* because it acts on the distribution function (which is a function of the *two* sets of variables q and p and therefore the analogue of a continuous matrix). However, L_{cl} is expressed in terms of a Poisson bracket (see equation 2.13) and is not factorizable.

Establishing a simple correspondence between classical and quantum superoperators will afford a source of insight into the structure of quantum mechanics.

Classical Commutation Rules

In a classical system with a single degree of freedom, four basic superoperators (two of which are multiplication superoperators) may be introduced:

$$Q, \quad P, \quad i\frac{\partial}{\partial P}, \quad -i\frac{\partial}{\partial Q} \qquad \text{(D.3)}$$

To emphasize that we consider them to be superoperators acting on distribution functions, we use uppercase letters. The multiplication by i is to obtain Hermitian superoperators.

Obviously these four quantities satisfy two independent noncommutation rules: one for P and $i(\partial/\partial P)$, the other for Q and $-i(\partial/\partial Q)$ (see the section on operators and complementarity in Chapter 3). In contrast, classical trajectory theory is entirely built in terms of functions of Q and P and does not admit any noncommutation relation.

Quantum theory therefore occupies an intermediate position because it leads to a *single* noncommutation relation between the quantum mechanical operators q_{op}, p_{op}. In this sense quantum mechanics is "more deterministic" than classical ensemble theory and less so than classical trajectory theory.

What is the meaning of the classical noncommutation rules? This question is studied in detail in a recent paper by Cl. George and myself.[1] As could be expected from the analogy with quantum mechanics, the following correspondence is obtained:

Superoperators	Eigendistribution
Q	Well-defined value of Q
P	Well-defined value of P
$-i\dfrac{\partial}{\partial Q}$	Uniform distribution in Q
$i\dfrac{\partial}{\partial P}$	Uniform distribution in P

The classical noncommutation rules therefore have a simple meaning: for example, a distribution function cannot simultaneously correspond to a well-defined value of Q and be independent of Q. The classical uncertainty relations therefore express a "logical" inconsistency. However, nothing prevents us from having, as an example, a distribution function that corresponds to well-defined values of both Q and P and thus to a classical trajectory.

Quantum Commutation Rules

Let us now introduce quantum mechanisms for four factorizable super-operators expressed in terms of the operators q_{op}, p_{op} that satisfy the Heisenberg uncertainty relation:

$$[p_{op}, q_{op}] = \frac{\hbar}{i} \tag{D.4}$$

These four superoperators are:

$$\tfrac{1}{2}(q_{op} \times I + I \times q_{op}), \quad \tfrac{1}{2}(p_{op} \times I + I \times p_{op}),$$

$$\frac{1}{h}(q_{op} \times I - I \times q_{op}), \quad \frac{1}{h}(p_{op} \times I - I \times p_{op}) \qquad \text{(D.5)}$$

There is a remarkable isomorphism between the classical superoperators (D.3) and the quantum superoperators (D.5). The commutation rules are identical and we may write the correspondence

$$Q \leftrightarrow \tfrac{1}{2}(q_{op} \times I + I \times q_{op})$$

$$i\frac{\partial}{\partial P} \leftrightarrow \frac{1}{h}(q_{op} \times I - I \times q_{op})$$

$$P \leftrightarrow \tfrac{1}{2}(p_{op} \times I + I \times p_{op})$$

$$-i\frac{\partial}{\partial Q} \leftrightarrow \frac{1}{h}(p_{op} \times I - I \times p_{op}) \qquad \text{(D.6)}$$

This correspondence permits us to attribute "similar" physical meaning to these sets of quantities. But that means that, using linear combinations and definition D.1, we have the correspondence

$$p_{op} \leftrightarrow P - i\frac{\hbar}{2}\frac{\partial}{\partial Q}, \quad q_{op} \leftrightarrow Q + i\frac{\hbar}{2}\frac{\partial}{\partial P} \qquad \text{(D.7)}$$

This result seems most interesting. The Hilbert space operators p_{op}, q_{op} cannot be expressed in terms of the quantities P, Q defined on a trajectory. They also involve superoperators acting on distribution functions. We now see clearly why the pure state of classical mechanics cannot be realized any longer in the Hilbert space: the coupling of the classical superoperators through the universal constant h prevents the realization of eigenensembles corresponding to well-defined values of both Q and P.

If we were to try to go from a continuous distribution function to a single point (a δ-function), the derivatives in expression D.7 would tend to infinity and we would obtain states of infinite energy. This precisely expresses the idea of correlations in phase space induced by h.

We see that the ensemble point of view, which has often been advocated in the past,[3] clarifies the position of quantum mechanics with respect to classical theory. It is not the appearance of noncommuting operators that is characteristic of quantum theory. This feature can always be incorporated in classical ensemble theory. The new and unique feature is the reduction of the four basic superoperators (expression D.5) in terms of the *two* combinations given in expression D.7. This is possible only through the existence of a universal constant having the physical dimensions of an action (momentum × coordinate). As a result, the concepts of momentum and coordinate in the Hilbert space are no longer independent and quantum theory seems an overdetermined classical theory in which the motion of neighboring points cannot be prescribed independently. Although there will never be a "classical" theory of quantum mechanics, a close analogy to the physical situation would be the classical motion of a string in which the motion of neighboring points can also no longer be prescribed independently—if we did so, this would lead to violent distortions of the string and to states of arbitrary high energy.

Concluding Remarks

As mentioned earlier in this appendix, we may also introduce noncommuting operators and a classical complementarity principle into the framework of classical ensemble theory. However, this principle has a trivial meaning: we cannot make contradictory statements about the distribution function ρ. The new feature in quantum mechanics is that the type of ensembles that can be constructed is limited by h. In addition, we can no longer take the limit to single trajectories and therefore the complementarity principle acquires a fundamental status in quantum mechanics.

It should be emphasized that at no point have we appealed to perturbations due to the observer or to other subjectivistic elements in this approach to quantum theory.

As in statistical mechanics, the transition from ensembles to trajectories is prevented by a change in the structure of the phase space. In statistical mechanics, it is the instability of motion that plays the critical role (see Chapter 9 and Appendixes A and C). Here the structure of the dynamic operators describing quantum ensembles leads to a theory that is both complete and probabilistic.

In conclusion, the difficult questions that were at the core of the celebrated Einstein-Bohr debate on the foundations of quantum theory (see *The Philosophy of Quantum Mechanics*[3]) begin to take new shapes: it is indeed possible to consider probabilistic theories that are complete and objective. Far from being the expression of ignorance, the probabilistic element may be the expression of new, fundamental features in the structure of the dynamic theory.

References

1. This discussion closely follows a recent paper by C. George and I. Prigogine, *Physica* 99A (1979): 369.
2. I. Prigogine, Cl. George, F. Henin, and L. Rosenfeld, *Chemica Scripta*, 4 (1973): 51.
3. For references to the work of Wigner, Moyal, Bopp, and others, see M. Jammer, *The Philosophy of Quantum Mechanics* (New York: Wiley, 1974), which includes an extensive bibliography.

REFERENCES

Allen, P. M. 1976. *Proc. Natl. Acad. Sci. U.S.* 73(3):665.

Allen, P. M., Deneubourg, J. L., Sanglier, M., Boon, F., and de Palma, A. 1977. Dynamic urban models. Reports to the Department of Transportation, under contracts TSC-1185 and TSC-1460.

Allen, P. M., and Sanglier, M. 1978. *J. Soc. Biol. Struct.* 1:265–280.

Arnold, L. 1973. *Stochastic differential equations.* New York: Wiley-Interscience.

Arnold, L., Horsthemke, W., and Lefever, R. 1978. *Z. Physik* B29:367.

Babloyantz, A., and Hiernaux, J. 1975. *Bull. Math. Biol.* 37:637.

Balescu, R. 1975. *Equilibrium and non-equilibrium statistical mechanics.* New York: Wiley-Interscience.

Balescu, R., and Brenig, L. 1971. Relativistic covariance of non-equilibrium statistical mechanics. *Physica* 54:504–521.

Barucha-Reid, A. T. 1960. *Elements of the theory of Markov processes and their applications.* New York: McGraw-Hill.

Bellemans, A., and Orban, J. 1967. *Phys. Letters* 24A:620.

Bergson, H. 1963. L'evolution créatrice. In *Oeuvres*, Editions du Centenaire. Paris: PUF.

Bergson, H. 1972. Durée et simultanéité. In *Mélanges*. Paris: PUF.

Bohr, N. 1928. *Atti Congr. Intern. Fis. Como, 1927*, vol. 2. [Also *Nature suppl.* (1928) 121:78.]

Bohr, N. 1948. *Dialectica* 2:312.

Boltzmann, L. 1872. *Wien. Ber.* 66:275.

Boltzmann, L. 1905. *Populäre Schriften*. Leipzig. (English translation published in 1974 by Reidel, Dordrecht/Boston.)

Bray, W. 1921. *J. Am. Chem. Soc.* 43:1262.

Briggs, T., and Rauscher, W. 1973. *J. Chem. Educ.* 50:496.

Caillois, R. 1976. Avant propos à la dissymétrie. In *Cohérences aventureuses*. Paris: Gallimard.

Chandrasekhar, S. 1943. *Rev. Mod. Phys.* 15(1).

Chandrasekhar, S. 1961. *Hydrodynamic and hydromagnetic stability*. Oxford: Clarendon.

Chapman, S., and Cowling, T. G. 1970. *Kinetic theory of non-uniform gases*. 3d ed. Cambridge University Press.

Clausius, R. 1865. *Ann. Phys.* 125:353.

Currie, D. G., Jordan, T. F., and Sudarshan, E. C. G. 1963. *Rev. Mod. Phys.* 35:350.

d'Alembert, J. 1754. Dimension, article in *l'Encyclopédie*, vol. N.

De Donder, Th. 1936. *L'affinité*. Revised edition by P. Van Rysselberghe. Paris: Gauthier-Villars.

d'Espagnat, B. 1976. *Conceptual foundations of quantum mechanics*. 2d ed. Reading, Massachusetts: Benjamin.

Dewel, G., Walgraef, D., and Borckmans, P. 1977. *Z. Physik.* B28:235.

Dirac, P. A. M. 1958. *The principles of quantum mechanics*. 4th ed. Oxford: Clarendon. (1st ed., 1930.)

Ehrenfest, P., and Ehrenfest, T. 1911. Begrifflische Grundlagen der Statistischen Auffassung der Mechanik. *Encyl. Math. Wiss.* 4:4. (English translation, *The conceptual foundations of statistical mechanics*, published in 1959 by Cornell University Press, Ithaca.)

Eigen, M., and Schuster, P. 1978. *Naturwissenschaften* 65:341.

Eigen, M., and Winkler, R. 1975. *Das Spiel*. München: Piper.

Einstein, A. 1917. Zum Quantensatz von Sommerfeld und Epstein. *Verhandl. Deut. Phys. Ges.* 19:82–92.

Einstein, A., and Besso, M. 1972. *Correspondence 1903–1955*. Paris: Hermann.

Einstein, A., and Born, M. 1969. *Correspondence 1916–1955*. Seuil. (See letter of September 7, 1944.)

Einstein, A., Lorentz, H. A., Weyl, H., and Minkowski, H. 1923. *The principle of relativity*. London: Methuen. (Dover edition.)

Erneux, T., and Hiernaux, J. In press.

Farquhar, I. E. 1964. *Ergodic theory in statistical mechanics*. New York: Interscience.

Feller, W. 1957. *An introduction to probability theory and its applications*, vol. 1. New York: Wiley.

Forster, D. 1975. *Hydrodynamic fluctuations, broken symmetry, and correlation functions*. New York: Benjamin.

George, Cl., Henin, F., Mayné, F., and Prigogine, I. 1978. New quantum rules for dissipative systems. *Hadronic J*. 1:520–573.

George, Cl., Prigogine, I., and Rosenfeld, L. 1973. The macroscopic level of quantum mechanics. *Kon. Danske Videns. Sels. Mat-fys. Meddelelsev* 38:12.

Gibbs, J. W. 1875–78. On the equilibrium of heterogeneous substances. *Trans. Connecticut Acad*. 3:108–248; 343–524. (See *Collected Papers*. New Haven: Yale University Press.)

Gibbs, J. W. 1902. *Elementary principles in statistical mechanics*. New Haven: Yale University Press. (Dover reprint.)

Glansdorff, P., and Prigogine, I. 1971. *Thermodynamic theory of structure, stability, and fluctuations*. New York: Wiley-Interscience.

Goldbeter, A., and Caplan, R. 1976. *Ann. Rev. Biophys. Bioeng*. 5:449.

Goldstein, H. 1950. *Classical mechanics*. Reading, Massachusetts: Addison-Wesley.

Golubitsky, M., and Schaeffer, D. 1979. An analysis of imperfect bifurcation. Proceedings of the Conference on Bifurcation Theory and Application in Scientific Disciplines. *Ann. N.Y. Acad. Sci*. 316:127–133.

Grecos, A., and Theodosopulu, M. 1976. On the theory of dissipative processes in quantum systems. *Acta Phys. Polon*. A50:749–765.

Haraway, D. J. 1976. *Crystals, fabrics, and fields*. New Haven: Yale University Press.

Heisenberg, W. 1925. *Z. Physik* 33:879.

Henon, M., and Heiles, C. 1964. *Astron. J*. 69:73.

Hirschfelder, J. O., Curtiss, C. F., and Bird, R. B. 1954. *The molecular theory of liquids*. New York: Wiley.

Hopf, E. 1942. *Ber. Math. Phys. Akad. Wiss*. (Leipzig) 94:1.

Horsthemke, W., and Malek-Mansour, M. 1976. *Z. Physik* B24:307.

Jammer, M. 1966. *Conceptual development of quantum mechanics*. New York: McGraw-Hill.

Jammer, M. 1974. *The philosophy of quantum mechanics*. New York: Wiley.

Kauffmann, S., Shymko, R., and Trabert, K. 1978. *Science* 199:259.

Kawakubo, T., Kabashima, S., and Tsuchiya, Y. 1978. *Progr. Theo. Phys*. 64:150.

Kolmogoroff, A. N. 1954. *Dokl. Akad. Nauk. USSR* 98:527.

Körös, E. 1978. In *Far from equilibrium*, ed. A. Pacault and C. Vidal. Berlin: Springer Verlag.

Koyré, A. 1968. *Etudes Newtoniennes*. Paris: Gallimard.

Lagrange, J. L. 1796. *Théorie des fonctions analytiques*. Paris: Imprimerie de la République.

Landau, L., and Lifschitz, E. M. 1960. *Quantum mechanics*. Oxford: Pergamon.

Landau, L., and Lifschitz, E. M. 1968. *Statistical physics*. 2d ed. Reading, Massachusetts: Addison-Wesley.

Leclerc, Ivor. 1958. *Whitehead's metaphysics*. London: Allen & Unwin.

Lefever, R., Herschkowitz-Kaufman, M., and Turner, J. W. 1977. *Phys. Letters* 60A:389.

Lemarchand, H., and Nicolis, G. 1976. *Physica* 82A:521.

McNeil, K. J., and Walls, D. F. 1974. *J. Statist. Phys.* 10:439.

Margalef, E. 1976. In *Séminaire d'écologie quantitative* (third session of E4, Venice).

Maxwell, J. C. 1867. *Phil. Trans. Roy. Soc.* 157:49.

May, R. M. 1974. *Model ecosystems*. Princeton, New Jersey: Princeton University Press.

Mehra, J., ed. 1973. *The physicist's conception of nature*. Dordrecht/Boston: Reidel.

Mehra, J. 1976. The birth of quantum mechanics. Conseil Européen pour la Recherche Nucléaire, 76-10.

Mehra, J. 1979. *The historical development of quantum theory: The discovery of quantum mechanics*. New York: Wiley-Interscience.

Minorski, N. 1962. *Nonlinear oscillations*. Princeton, New Jersey: Van Nostrand.

Misra, B. 1978. *Proc. Natl. Acad. Sci. U.S.* 75:1629.

Misra, B., and Courbage, M. In press.

Monod, J. 1970. *Le hasard et la nécessité*. Paris: Seuil. (English translation, *Chance and necessity*, published in 1972 by Collins, London.)

Morin, E. 1977. *La méthode*. Paris: Seuil.

Moscovici, S. 1977. *Essai sur l'histoire humaine de la nature*. Collection Champs Philosophique. Paris: Flammarion.

Moser, J. 1974. *Stable and random motions in dynamical systems*. Princeton, New Jersey: Princeton University Press.

Nicolis, J., and Benrubi, M. 1976. *J. Theo. Biol.* 58:76.

Nicolis, G., and Malek-Mansour, M. 1978. *Progr. Theo. Phys. suppl.* 64:249–268.

Nicolis, G., and Prigogine, I. 1971. *Proc. Natl. Acad. Sci. U.S.* 68:2102.

Nicolis, G., and Prigogine, I. 1977. *Self-organization in non-equilibrium systems*. New York: Wiley.

Nicolis, G., and Prigogine, I. In press. Non-equilibrium phase transitions. *Sci. Am.*

Nicolis, G., and Turner, J. 1977a. *Ann. N.Y. Acad. Sci.* 316:251.

Nicolis, G., and Turner, J. 1977b. *Physica* 89A:326.

Noyes, R. M., and Field, R. J. 1974. *Ann. Rev. Phys. Chem.* 25:95.

Onsager, L. 1931a. *Phys. Rev.* 37:405.

Onsager, L. 1931b. *Phys. Rev.* 38:2265.

Pacault, A., de Kepper, P., and Hanusse, P. 1975. *C. R. Acad. Sci. (Paris)* 280C:197.

Paley, R., and Wiener, N. 1934. *Fourier transforms in the complex domain.* Providence, Rhode Island: American Mathematical Society.

Planck, M. 1930. *Vorlesungen über Thermodynamik.* Leipzig. (English translation, Dover.)

Poincaré, H. 1889. *C. R. Acad. Sci. (Paris)* 108:550.

Poincaré, H. 1893a. Le mécanisme et l'expérience. *Rev. Metaphys.* 1:537.

Poincaré, H. 1893b. *Les méthodes nouvelles de la mécanique céleste.* Paris: Gauthier-Villars. (Dover edition, 1957.)

Poincaré, H. 1914. *Science et méthode.* Paris: Flammardon.

Poincaré, H. 1921. Science and hypothesis. In *The foundations of science.* New York: The Science Press.

Popper, K. 1972. *Logic of scientific discovery.* London: Hutchinson.

Prigogine, I. 1945. *Acad. Roy. Belg., Bull. Classe Sci.* 31:600.

Prigogine, I. 1962a. *Introduction to nonequilibrium thermodynamics.* New York: Wiley-Interscience.

Prigogine, I. 1962b. *Nonequilibrium statistical mechanics.* New York: Wiley.

Prigogine, I. 1975. Physique et métaphysique. In *Connaissance scientifique et philosophie.* Publication no. 4 of the Bicentennial, Royal Academy of Belgium.

Prigogine, I. Forthcoming. *The microscopic theory of irreversible processes.* New York: Wiley.

Prigogine, I., and George, C. 1977. New quantization rules for dissipative systems. *Intern. J. Quantum Chem.* 12(suppl. 1):177–184.

Prigogine, I., George, C., Henin, F., and Rosenfeld, L. 1973. A unified formulation of dynamics and thermodynamics. *Chem. Scripta* 4:5–32.

Prigogine, I., and Glansdorff, P. 1971. *Acad. Roy. Belg., Bull. Classe Sci.* 59:672–702.

Prigogine, I., and Grecos, A. 1979. Topics in nonequilibrium statistical mechanics. In *Problems in the foundations of physics.* Varenna: International School of Physics "Enrico Fermi."

Prigogine, I., Herman, R., and Allen, P. 1977. The evolution of complexity and the laws of nature. In *Goals in a global community: A report to the Club of Rome,* vol. 1, ed. E. Laszlo and J. Bierman. Oxford: Pergamon.

Prigogine, I., Mayne, F., George, C., and De Haan, M. 1977. Microscopic theory of irreversible processes. *Proc. Natl. Acad. Sci. U.S.* 74:4152–4156.

Prigogine, I., and Stengers, I. 1977. The new alliance, parts 1 and 2. *Scientia* 112:319–332; 643–653.

Prigogine, I., and Stengers, I. 1979. *La nouvelle alliance.* Paris: Gallimard.

Prigogine, I., and Stengers, I. Forthcoming. *Science and metascience.* New York: Doubleday.

Rice, S., Freed, K. F., and Light, J. C., eds. 1972. *Statistical mechanics: New concepts, new problems, new applications.* Chicago: University of Chicago Press.

Rosenfeld, L. 1965. *Progr. Theoret. Phys. Suppl.* Commemoration issue, p. 222.

Ross, W. D. 1955. *Aristotle's physics.* Oxford: Clarendon.

Sambursky, S. 1963. *The physical world of the Greeks.* Translated from the Hebrew by M. Dagut. London: Routledge and Kegan Paul.

Schlögl, F. 1971. *Z. Physik.* 248:446.

Schlögl, F. 1972. *Z. Physik.* 253:147.

Schrödinger, E. 1929. Inaugural lecture (Antrittsrede), 4 July 1929. (English translation in *Science, theory, and men,* published in 1957 by Dover.)

Serres, M. 1977. *La naissance de la physique dans le texte de Lucréce: Fleuves et turbulences.* Paris: Minuit.

Sharma, K., and Noyes, R. 1976. *J. Am. Chem. Soc.* 98:4345.

Snow, C. P. 1964. *The two cultures and a second look.* Cambridge University Press.

Spencer, H. 1870. *First principles.* London: Kegan Paul.

Stanley, H. E. 1971. *Introduction to phase transitions and critical phenomena.* Oxford University Press.

Theodosopulu, M., Grecos, A., and Prigogine, I. 1978. *Proc. Natl. Acad. Sci. U.S.* 75:1632.

Theodosopulu, M., and Grecos, A. 1979. *Physica* 95A:35.

Thom, R. 1975. *Structural stability and morphogenesis.* Reading, Massachusetts: Benjamin.

Thomson, W. 1852. *Phil. Mag.* 4:304.

Tolman, R. C. 1938. *The principles of statistical mechanics.* Oxford University Press.

Turing, A. M. 1952. *Phil. Trans. Roy. Soc. London, Ser. B.* 237:37.

Von Neumann, J. 1955. *Mathematical foundations of quantum mechanics.* Princeton, New Jersey: Princeton University Press.

Welch, R. 1977. *Progr. Biophys. Mol. Biol.* 32:103–191.

Whittaker, E. T. 1937. *A treatise on the analytical dynamics of particles and rigid bodies.* 4th ed. Cambridge University Press. (Reprint, 1965.)

NAME INDEX

SUBJECT INDEX